The Sustainable Forestry Handbook

SECOND EDITION

The Sustainable Forestry Handbook

A practical guide for tropical forest managers on implementing new standards

SECOND EDITION

*Sophie Higman, James Mayers, Stephen Bass, Neil Judd
and Ruth Nussbaum*

ProForest

iied

International
Institute for
Environment and
Development

London • Sterling, VA

This publication is an output from a research project funded by the United Kingdom Department for International Development (DFID) for the benefit of developing countries. The views expressed are not necessarily those of DFID. R6370 – Forestry Research Programme.

First published in the UK and USA
Earthscan Publications Ltd

First South Asian Edition 2006

ISBN: 1-84407-118-9

Typesetting by Mapset Ltd, Gateshead, UK
Printed and bound by Replika Press Pvt. Ltd., India
Cover design by Ruth Bateman
Cover photo © Sophie Higman
Illustrations © Kathryn K Davis
Icons © Christine Bass

For full list of publications please contact:

Earthscan
8–12 Camden High Street
London, NW1 0JH, UK
Tel: +44 (0)20 7387 8558
Fax: +44 (0)20 7387 8998
Email: earthinfo@earthscan.co.uk
Web: **www.earthscan.co.uk**

22883 Quicksilver Drive, Sterling, VA 20166-2012, USA

Earthscan is an imprint of James and James (Science Publishers) Ltd and publishes is association with the International Institute for Environment and Development

A catalogue record for this book is available from the British Library

Library of Congress Cataloging-in-Publication Data has been applied for

This edition is for sale in India, Pakistan, Bangladesh, Sri Lanka and Nepal only. Not for export elsewhere.

Contents

List of Tables, Figures and Boxes

Tables

Figures

Boxes

Acronyms and Abbreviations

AAC	annual allowable cut
ACT	Amazon Co-operation Treaty
ATO	African Timber Organisation
BOP	best operating practice
CAR	corrective action request
CATIE	Centro Agronomico Tropical de Investigación y Enseñanza
CBA	cost-benefit analysis
CIFOR	Center for International Forestry Research
CITES	Convention on International Trade in Endangered Speices
CSA	Canadian Standards Association
CSD	UN Commission on Sustainable Development
dbh	diameter at breast height
EIA	Environmental Impact Assessment
EMAS	European Union Environmental Management and Auditing System
EMP	Environmental Management Programme
EMS	Environmental Management System
ESIA	Environmental and Social Impact Assessment
FAO	UN Food and Agriculture Organization
FMP	forest management plan
FMU	forest management unit
FORIG	Forest Research Institute of Ghana
FSC	Forest Stewardship Council
FSC P&C	Forest Stewardship Council Principles and Criteria for forest management
FTN	Forest and Trade Network
GFTN	Global Forest and Trade Network (of WWF)
HCVF	High Conservation Value Forest
IIED	International Institute for Environment and Development
ILO	International Labour Organization
IPF	Inter-governmental Panel on Forests
IPM	integrated pest management
ISO	International Organization for Standardization
ITTA	International Tropical Timber Agreement
ITTO	International Tropical Timber Organization
IUCN	The World Conservation Union
LEI	Lembaga Ekolabel Indonesia
MIV	Modular Implementation and Verification
MTCC	Malaysian Timber Certification Council
NGO	non-governmental organization
NTFP	non-timber forest product
PAFCS	Pan-African Forest Certification Scheme
PEFC	Programme for the Endorsement of Forest Certification Schemes
PEOLG	Pan-European Operational Level Guidelines
PSPs	permanent sample plots

SATGA	South African Timber Growers Association
SFB	Sustainable Forestry Board
SFI	Sustainable Forestry Initiative
SFM	sustainable forest management
SLIMF	Small and Low Intensity Managed Forests
SIA	Social Impact Assessment
UNCED	United Nations Conference on Environment and Development
VJR	virgin jungle reserve
WCED	World Commission on Environment and Development
WWF	World Wide Fund for Nature

Acknowledgements

Many people contributed to the first edition of *The Sustainable Forestry Handbook*. Editorial assistance was provided by Nicola Baird. Particular help came from the reviewers of the first draft, whose contributions were invaluable. However, the views expressed in *The Sustainable Forestry Handbook* are those of the authors alone. Thanks to: Prof Marielos Alfaro, Escuela Ciencias Ambientales, Costa Rica; George K Bruce, Eastern Highlands Plantations Ltd, Zimbabwe; Prof E F Bruenig DSc, University of Hamburg, Germany; Froylan Castaneda, Food and Agriculture Organization, Italy; Dr Jane Clark, Department for International Development, UK; Carol J Pierce Colfer, Center for International Forestry Research (CIFOR), Indonesia; Keith Dolman, Ghana Forest Sector Development Project, Ghana; Cathleen Fogel, University of California, US; Dr Alastair I Fraser, UK Tropical Forest Management Programme, Indonesia; Peter C Gondo, Forestry Commission, Zimbabwe; Steve Gretzinger, Rogue Institute for Ecology and Economy, US; Guyana Forestry Commission Support Project, Guyana; Dr Michael Kleine, Malaysian–German Sustainable Forest Management Project, Malaysia; Michael Jourdain, forestry consultant, UK; Bas Louman, Centro Agronomico Tropical de Investigación y Enseñanza (CATIE), Costa Rica; Stewart Maginnis, Proyecto de Manejo Integrado del Bosque Natural, Costa Rica; Gerrit Marais, South African Forestry Company Ltd, South Africa; Bill Maynard, independent consultant, UK; Ashley R Parasram, European Forest Institute, Finland; Simon Rietbergen, The World Conservation Union (IUCN), Switzerland; Jim Sandom, Woodmark Scheme and Responsible Forestry Programme, The Soil Association, UK; Dr Jaap Schep, adviser to the Solomon Western Islands Fair Trade programme (SWIFT), The Netherlands; Dr John S B Scotcher, Sappi Forests Pty Ltd, South Africa; Hannah Scrase, Forest Stewardship Council (UK), UK; J W Heezen, Skal Forestry Certification, The Netherlands; Prof Virgílio Maurício Viana, University of São Paulo, Brazil; Toni Williams, A Mazaharally and Sons Ltd, Guyana.

Also many thanks to Steve Jennings, Nilofer Ghaffar and Ian Gray for their help with the second edition.

This handbook is not an official FSC or ITTO publication.

Using This Handbook

The Sustainable Forestry Handbook provides forest managers with the necessary tools to understand and put into practice new standards for forest management. It is based around two major international initiatives for the promotion of sustainable forest management, which were developed by the Forest Stewardship Council (FSC) and the International Tropical Timber Organization (ITTO). These are the main approaches which have been promoted, especially in the tropics, by recent market and policy initiatives. The Sustainable Forestry Handbook also provides an environmental management system framework, based on ISO 14001, to assist implementation of the standards.

The handbook provides a framework for understanding, planning and implementing improved forest management techniques. It gives a method for identifying what activities are necessary, provides guidance on how to approach them and points the way to further sources of information. Case studies describe examples of existing situations; common problems are highlighted and some possible solutions suggested.

The Sustainable Forestry Handbook is ideal for use by forest managers and forest management teams in tropical, developing countries where up-to-date information and guidance can be hard to find. It aims to provide basic and comprehensive guidance on practical methods in meeting international standards. It will be of particular help for forest organizations which are seeking independent certification of their forest management.

Some of the methods described in The Sustainable Forestry Handbook, especially in the social areas, may be new to forest managers and even to some researchers. They are included because it is important for the forest manager to be aware of current thinking about sustainable forest management

Introduction to the Second Edition

Since the first edition of The Sustainable Forestry Handbook was published in 1999, standards in forestry have moved forward, new standards have been developed and existing standards modified. Part One of the second edition reflects these changes and gives an overview of current initiatives and the context for implementing sustainable forest management.

However, this second edition maintains its focus on the practical application of the ITTO Guidelines and FSC standards, which are still the main international standards applicable to tropical forests. Significant updates are included relating to the treatment of high conservation value forests, social issues covered by International Labour Organization (ILO) conventions and forest management certification.

What is in this Handbook?

The Sustainable Forestry Handbook provides an introduction to international standards for forest management. What are they? What do they require? How can it be done? The handbook is divided into six main parts, followed by a number of appendices:

PART ONE: INTRODUCTION TO INTERNATIONAL STANDARDS FOR FOREST MANAGEMENT

This provides an introduction to sustainable forest management and an overview of the main international and national initiatives attempting to define sustainable forest management. The conditions that enable forest managers to adopt sustainable forest management are discussed.

PART TWO: WHAT DO THE STANDARDS REQUIRE?

Part Two describes what is required, as defined by two of the major international initiatives – FSC and ITTO. It covers legal and policy issues; sustained timber production; environmental protection; the wellbeing of people and a section specifically covering plantation management.

PART THREE: USING AN ENVIRONMENTAL MANAGEMENT SYSTEM

This part provides a *management framework* for sustainable forest management activities. It is loosely based on the structure of the internationally applied environmental management system, ISO 14001. Cross-linkages to forest management activities are highlighted.

PART FOUR: MEETING THE REQUIREMENTS

Part Four describes *how* the requirements described in Part Two might be met, in terms of planning, implementation and monitoring. This incorporates the major forest management activities which need to be addressed.

PART FIVE: TACKLING SOCIAL ISSUES

Sustainable forest management involves social skills as much as it does silvicultural know-how. This special focus on social issues helps bring forest managers up to date with ways to include all forest stakeholders in the management process. New techniques successfully used in community forest projects and agriculture are described.

PART SIX: FOREST MANAGEMENT CERTIFICATION

Independent certification of forest management is discussed in more detail. Brief consideration is given to the reasons why an organization might want to become certified, which standards can be used for certification and the process leading to the award of a certificate.

Following the main text of the handbook is a selection of further information:
- The *Appendices* contain additional background information, references, examples and contact details.
- The *Glossary* explains some unfamiliar or confusing terms.
- An *Index* identifies the most common themes.

How to Use this Handbook

This handbook is not designed to be read from cover to cover! It is a reference book to be consulted as necessary on specific subjects. There is, therefore, some repetition between the sections. The handbook is designed to be used by forest managers in very different situations. Each of the six parts describes different aspects of implementation of sustainable forest management. Some aspects are more appropriate in some situations than others.

A number of ways of navigating around this handbook are provided:

- The table of contents lists overall subject areas.
- The index provides more precise locations for specific themes.
- The icons in the top right corner of the pages identify sections at a glance.
- The cross-referencing leads from one topic of interest to other connected areas.
- The list of tables, figures and boxes shows all the areas picked out for special attention.

Part One

Introduction to International
Standards for Forest Management

Introduction to Part One

There is increasing pressure worldwide for improvement in the quality of forest management. Concern about environmental and social issues associated with forestry – such as effects on biodiversity, climate change, desertification, flooding, conflicts over use rights and sustainable development generally – has led to international agreements and programmes for improving forest management practices.

Although there is general agreement that sustainable forest management (SFM) should be environmentally responsible, socially beneficial and economically viable, a need to agree a more precise definition of SFM has been recognized. As a result, various attempts have been made to develop international and national standards of sustainable forest management. However, it is often difficult for forest managers, especially in the tropics, to find practical information explaining exactly what is required and how to put it into practice. This handbook aims to fill that gap.

Part One looks briefly at the background to SFM and the development of new standards of forest management. It covers:

 Chapter 1 What is Sustainable Forest Management?

 Chapter 2 Why practise Sustainable Forestry Management?

 Chapter 3 Standards for Sustainable Forest Management

1 What is Sustainable Forest Management?

Sustainable forest management has been described as forestry's contribution to sustainable development. This is development which is economically viable, environmentally benign and socially beneficial, and which balances present and future needs (see Box 1.1).

Interactions between the flora and fauna in a forest ecosystem are complex and often poorly understood. The consequences of actions taken today may only show up in 50 to 100 years' time. For this reason, some people feel that the phrase 'sustainable forest management' should not be used to describe current management systems. Other phrases such as 'good forest stewardship' or 'well-managed forests' are often preferred.

Although attempts at comprehensive definitions of SFM will always be argued over, there are many practices that are widely recognized as unsustainable, and which inevitably lead to forest degradation. For example, actions which cause a breakdown in the forest's ability to regenerate, or that exclude local people's needs, are unsustainable.

The approach taken in this handbook is to define practical sustainable forest management as the best available practices, based on current scientific and traditional knowledge, which allow multiple objectives and needs to be met without degrading the forest resource.

Early definitions of sustainable forestry concentrated on the timber resource, with management aimed at the 'sustained yield' of a limited number of wood products. Recently the importance of other products and services provided by forests has been recognized, particularly those of broader social concern. Concepts of SFM now encompass the continued production of these, such as protection of water supply, soils and cultural sites, as well as timber.

Box 1.1   What does sustainable forest management mean?

One of the most widely accepted definitions of sustainable development was produced by the World Commission on Environment and Development in 1987. This defined sustainable development as:

'*Development that meets the needs of the present without compromising the ability of future generations to meet their own needs.*'

The concept of sustainable development recognizes that utilization will change natural ecosystems, but that conservation is also important. It also recognizes that utilization of forests is important for achieving social goals, such as poverty alleviation. These aims have to be balanced. Most governments now adhere to the concept of sustainable development and incorporate it in new policies.

There are various definitions of sustainable forest management, but they all say essentially the same:

'*Sustainable forest management is the process of managing forests to achieve one or more clearly specified objectives of management with regard to the production of a continuous flow of desired forest products and services, without undue reduction of its inherent values and future productivity and without undue undesirable effects on the physical and social environment.*'

(*ITTO, Criteria and Indicators for Sustainable Management of Natural Tropical Forests, 1998*)

Management of a forest for a single product will affect the forest's ability to provide other services or products, so trade-offs have to be made.

For example, managing the forest for high levels of timber production may affect the value of the forest as a habitat for wild animals. It is not possible to maximize production of everything, all the time. This means forest managers applying SFM must define the balance of different management objectives that they are aiming to achieve. It also means that the objectives of forest management will change over time, as different forest products and services become more valued, or less desirable, and as we learn more about what the forest can sustain.

1.1 Elements of Sustainable Forest Management

As described later in Part One, there are a multitude of initiatives to define the major components of sustainable forest management in practice. However, most initiatives have in common the elements shown below and the remainder of this handbook is based around these elements.

1 A legal and policy framework
   • compliance with legislation and regulation;
   • tenure and use rights;
   • the forest organization's commitment and policy.
2 Sustained and optimal production of forest products
   • management planning;
   • sustained yield of forest products;
   • monitoring the effects of management;
   • protection of the forest from illegal activities;
   • economic viability and optimizing benefits from the forest.
3 Protecting the environment
   • environmental impact assessment;
   • conservation of biodiversity;
   • ecological sustainability;
   • use of chemicals;
   • waste management.
4 Wellbeing of people
   • consultation and participation processes;
   • social impact assessment;
   • recognition of rights and culture;
   • relations with employees;
   • contribution to development.
5 Some extra considerations apply specifically to plantations and focus on:
   • plantation planning;
   • species selection;
   • soil and site management;
   • pest and disease management;
   • conservation and restoration of natural forest cover.

2 Why practise Sustainable Forest Management?

Why should forest managers improve their practices and implement the requirements of sustainable forest management described in this book? What are the conditions that encourage or require forest managers to adopt sustainable forest management practices, and what constraints do they face in improving their forest practices? This chapter addresses these questions.

2.1 Forest Governance

FOREST GOVERNANCE: WHAT IS IT AND WHY DOES IT MATTER?

Forest managers operate within a framework of laws, policies and institutional processes – they are 'governed'. Forest governance requirements are therefore a fundamental factor affecting the forest manager's decisions on whether to work towards sustainable forest management.

Forest governance is about the policy, legal and institutional conditions that affect how people treat forests. It generally refers to the quality of decision-making processes – their transparency, accountability and equity – rather than the formal political structures of government. Good forest governance supports and encourages the implementation of sustainable forest management (SFM). At the same time, forest managers who implement SFM can themselves help bring about better forest governance.

Forest governance spans local to global levels. Governance pressures to implement SFM may be exerted at a number of levels, including:

Box 2.1   What is governance?

Governance is a notion commonly held to be very close to that of 'government' – more or less 'what governments do'. But over the last decade in particular, governance as a term has become commonly used in a range of contexts, such as corporate governance, international governance, national governance and local governance. It is often now used in a general sense to mean the process of decision-making and the process by which decisions are implemented (or not implemented). One useful definition of governance is: 'the traditions, institutions and processes that determine how power is exercised, how citizens are given a voice, and how decisions are made on issues of public concern'.[1]

Governments are not the only organizations involved in governance. These days we are all told what to do, not only by our national or local governments, we are 'governed' by a range of non-governmental organizations too, some of them more accountable to us than others. At the national level, informal decision-making structures, such as groups of special advisers may exist. More locally, powerful families or companies may make or influence decisions that affect us.

---

1 Institute on Governance, 2004, http://www.iog.ca/

- the local level (eg community rules and social norms regarding forest use);
- the national level (eg legal rights to forest land and resources, and policies affecting the relative profitability of different forest uses);
- the global level (eg multilateral environmental agreements affecting forests, trade rules, and the policies of multinational companies and investors), there is an increasing trend for global corporations to exert influence on local levels.

Forest governance is further complicated by multi-stakeholder and multi-sectoral interactions. Forests are resources that a wide variety of groups use in different ways, for specific goods and services, or for

---

Box 2.2   ## Some qualities of good governance and how forest managers can contribute

Good governance is about good decisions over matters of public concern. This may sound grand and idealistic, but achieving it requires many small practical steps to be taken – and forest managers can take some of them. Forest managers – linked as they often are to a range of other groups and to significant areas of land and resources – can thus play a major role in developing good governance.

Rule of law

Fair legal frameworks that are impartially enforced are needed. Forest managers should obey the law, engage in dialogue about its inconsistencies where necessary, avoid corrupt practices and encourage others with whom they interact to do likewise.

Transparency

Decisions need to be taken in accordance with agreed rules. Relevant information must be accessible to those who will be affected. Forest managers can help publicize decision-making processes (such as standards development processes) and communicate information that might affect decisions, in ways that are easily understandable to forest users, neighbouring communities, workers and other organizations.

Equity

All citizens should feel they have a stake in, and are not excluded from, society. This requires that all groups, but particularly the most vulnerable, have opportunities to maintain or improve their wellbeing. Forest managers can ensure that their own objectives are clearly expressed and that people with rights or interests in forest areas under their control receive fair treatment.

Efficiency

Citizens and institutions should produce results that meet the needs of society while making the best use of resources at their disposal. Forest managers are key players in ensuring that forest goods and services are used productively and sustainably.

Accountability

Governmental institutions, private sector enterprises and civil society organizations should be answerable to those who are affected by their actions. Forest managers should themselves be accountable and press for accountability in the organizations with which they interact.

    The above qualities of good governance depend on and reinforce each other: for example, accountability depends on transparency, while equity is reinforced by the rule of law.

---

conversion into other forms of capital (eg cash and deforested land for farming or human settlement). The relative profitability of different forms of forest use and conversion is, therefore, a significant determinant of whether SFM will be implemented.

Values and structures of government influence forest governance. Decisions on forests are influenced by the values of the decision-makers, but these values can vary widely. The structure of government will also affect how governance is exercised. For example, federal systems tend to operate differently from centralized systems in their control of forest management. Other aspects such as history, ecology and economic conditions will also influence governance.

When forest governance is good it sets clear boundaries and conditions within which society gets what it wants, and is prepared to pay for, from its forests. But, where governance is bad, forest managers face some tricky decisions about how to operate. Box 2.2 shows some of the qualities of good governance and how forest managers can contribute to making it more widespread.

GOOD FOREST GOVERNANCE DEMANDS SFM

One reason why forest managers may need to practise sustainable forest management is because forest governance tells them to do so. Forest governance has changed in many countries in recent years, and increasingly promotes policies, laws and institutions that are conducive to sustainable forest management. For example:

- policy debates and implementation tend now to involve a range of groups and partnerships, not merely government and some elites as they used to;
- policy objectives in many countries have opened up, from overriding concerns with forests as timber resources or land for development, to a concern for a wider range of forest goods and services;
- forest-dependent communities in some countries are beginning to have their rights recognized, enabling them to be more effective forest managers;
- a number of international programmes aim to improve governance of the sector.

Recent years have seen some progress in understanding the components of good forest governance and how they fit together to encourage or require SFM. It has also become clear that forest managers themselves have a key role to play in installing the components of good forest governance.[2] Although forest governance varies widely from one context to another, it is possible to identify some components that are common to a wide range of different countries. These can be represented as tiers in a simple pyramid as shown in Figure 2.1.[3]

The pyramid aims to highlight the idea that:

- Good forest governance can be 'built' – it is made up of basic building blocks put in place by ordinary human beings, including forest managers! Forest managers can use their knowledge and influence to help in establishing the foundations, such as the basic framework of rights and investment conditions, right through to negotiating the mix of policies and instruments used, and participating in the verification of good practice.

---

2 A multi-country analysis of 'what works' (and what fails) in forest policy and institutional process has been developed in: J Mayers and S Bass, *Policy That Works for Forests and People.* Earthscan, London, 1999, www.iied.org/forestry/pubs/ptw.html#overviewreport

3 J Mayers, S Bass and D Macqueen, *The Pyramid: A Diagnostic and Planning Tool for Good Forest Governance.* IIED, London and the World Bank WWF Alliance for forest conservation and sustainable use, Washington DC, 2002, http://www.iied.org/docs/flu/PT7_pyramid.pdf

5 **Verification of SFM.** Audit, certification
or participatory review undertaken

4 **Extension.** Promotion of SFM to consumers and
all those linked to forests undertaken

3 **Instruments.** Coherent set of 'carrots and sticks' for
implementation in place

2 **Policies.** Forest policies, standards for SFM and legislation in place

1 **Roles.** Institutional roles in forestry and land use negotiated and developed

FOUNDATIONS
Property/tenure rights and constitutional guarantees
Market and investment conditions
Mechanisms for engagement with extra-sectoral influences
Recognition of lead forest institutions (in government, civil society and private sector)

FIGURE 2.1 The 'pyramid' of good forest governance

- Some elements of forest governance are more fundamental than others. The lower tiers in the pyramid take more building than the upper tiers: there are more elements involved and they require a wide range of others to play their part alongside forest managers.
- Some elements of forest governance are prerequisites for others. The vertical arrangement of the tiers suggests a sequence, but some 'gravity-defying' progress can be made in reality on upper tiers even when lower ones are not complete.
- Depending on the context, forest managers will be able and willing to tackle some elements of forest governance more than others. It is important that the 'big picture' be kept in mind – of the connections between elements – so that pitfalls are avoided and opportunities seized. For example, forest managers who promote the merits of SFM to others might kick off a wider process of questioning in the forestry sector leading to a useful re-negotiation and sorting out of the key institutional roles for SFM.

In short, good governance requires SFM and supports it by providing solid foundations, frameworks of policies and institutions, and the mechanisms for spreading and verifying good practice. SFM works best in places where good governance is in place. However, nowhere is perfect and there are usually ways in which forest managers can operate to improve the elements of forest governance themselves.

TACKLING BAD FOREST GOVERNANCE NEEDS GOOD SFM MODELS

Where forest governance is bad – where policies and institutions obstruct and undermine the prospects for SFM – forest managers who can demonstrate the benefits of SFM are needed more than ever. By practising SFM, forest managers can help bring about better forest governance.

It is important that SFM initiatives, such as demonstrations of good practice at the level of the forest management unit or processes for developing standards, engage with governance issues. Many do, for example:

- SFM initiatives often bring key decision-makers together with local practitioners, encouraging a shared understanding about good forestry, how to assess it and who should be responsible.
- Examples of SFM at the forest level are important to show what can be achieved, even within existing governance constraints.
- SFM initiatives can improve recognition of the rights and potentials of local forest groups.
- Where ownership or control is being transferred under decentralization or privatization programmes, the capacity-building that SFM initiatives provide can be effective entry points for improving governance.

In short, situations with bad forest governance need examples of SFM – to maintain viability, to show what is possible and to inspire others. Forest managers can work with policy-makers and others within their capabilities to tackle some of the constraints of bad governance and to promote SFM more widely.

OTHER ENABLING CONDITIONS FOR SFM

Good governance is essential to create the right environment for sustainable forest management in the long term. However, SFM will still not occur if no one knows how to manage the forest in practice. The technical capacity for SFM has increased significantly over the last decade, particularly incorporating a better understanding of environmental and social aspects of forest management.

However, in some countries, practical, technical constraints still exist on forest managers' ability to implement SFM practices. To implement SFM, forest managers also need:

- **Knowledge about SFM:** our understanding of the components of SFM in practice has changed rapidly over the past decade. Key stakeholders involved in forest management, including forest managers, government, non-governmental organizations (NGOs) and other stakeholders need to understand what SFM means in order to work together to achieve it.
- **Information about the forest resource:** forest management must be based on a reasonable understanding of the forest resource and its response to forest management.
- **Management techniques:** traditionally forest management has focused on timber production. As management objectives increasingly include other products and services, such as non-timber forest products, water or carbon, good management practices need to be developed to produce them. These practices are currently not always known or documented.
- **Understanding the environment and conservation:** many forest managers understand the ideas of minimizing environmental disturbance during logging operations, but may have little knowledge of how to plan, implement and monitor programmes to protect biodiversity and wider environmental values.
- **Consultation and working with stakeholders:** definitions and standards for SFM generally incorporate consultation and collaboration with a range of stakeholders as an important element of sustainable management. Most forest managers are untrained and unskilled in this process, while specialist help may also be hard to find.
- **Training:** appropriate training courses for all levels of staff working in forest management are essential to ensure that the requirements of SFM are actually implemented on the ground.
- **Resources:** implementing SFM can have high initial costs in terms of both human and financial resources. Implementing new practices and purchasing new machinery can be expensive initially. In addition, forest managers are expected to deliver a wider range of services and benefits to society as part of SFM, while these are not usually directly paid for.

Good forest governance provides the framework that encourages forest managers to adopt sustainable forest management practices. Ensuring that the technical capacity to implement SFM is available in a country is also essential for creating the conditions that allow improved forest management on the ground.

THE NET IS CLOSING IN ON ILLEGAL FORESTRY

When there is a total failure of forest governance, endemic illegal logging is often the result. Illegal logging and the associated trade is now recognized as affecting developing and industrialized countries, tropical and temperate forests, alike. Countries that import illegal timber sustain the demand for these illegal products. The World Bank estimates that illegal logging results in US$10–15 billion per year in lost revenue to governments.

Illegal logging impacts directly on poor people by denying them contributions to their livelihoods and by degrading their environment. It is also often associated with corruption, ranging from give-away logging concessions to fraud and tax evasion. More widely, illegal logging undermines the prospects for sustainability in forestry. It is difficult for timber, produced through good forest management practices, to compete economically with the products of illegal logging.

No single action can stop illegal logging. Control requires the implementation of a range of measures to identify and reduce the consumption of illegal products in countries that import timber, as well as simultaneous measures to strengthen the enforcement of legal controls on forest management and logging in timber-producing countries.[4]

The enforcement of legal requirements is not always as simple as it sounds. Enforcement of current laws is in some contexts irrelevant, or even detrimental to poor people. Laws sometimes prop up existing exploitation systems, denying the rights of poor people at a local level. Legality cannot be enforced without regard for human rights and livelihood opportunities.[5]

Significant international efforts have emerged to combat illegal logging and the power of unscrupulous forest industries to trade illegally. These include the inter-governmental processes in East Asia, Europe and Africa on forest law enforcement, governance and trade. The aim of these processes is to strengthen controls on illegal logging, and improve the verification and identification of the products of legal forest operations. In consuming countries, public and private enterprise procurement policies that exclude illegal products are being developed. At the same time, such initiatives place more emphasis on encouraging trade in products of SFM. International trade in illegal forest products is likely to become more difficult in future.

2.2 Environmental and Social Risk Reduction

GROWING PUBLIC AWARENESS OF FORESTRY'S IMPACTS

Forest managers take precautions to minimize risks that might affect their forest (such as fire or pest damage) and prepare contingency plans to deal with situations that may arise. The concept of 'environmental and social risk' reflects a growing awareness about environmental and social impacts and their consequences. Poor forest management is an area that has attracted particular public and governmental concern and, as a result, forest organizations have a relatively high level of environmental and social risk attached to them.

Forest issues have become such major concerns in many countries because large areas of land are often involved, large amounts of money may be made and many people's livelihoods are connected to forests. Forests have often formed the power base for many governments and social groups. Finally, forest issues can be highly contentious because specific groups of people understand and value them very differently.

Heightened public awareness of the impacts of forestry means that forest managers need to take steps to improve practice where necessary and to demonstrate that progress is being made. A forest

4 D Brack, K Gray and G Hayman, *Controlling the International Trade in Illegally Logged Timber and Wood Products*. Royal Institute of International Affairs, London, 2002, www.riia.org/pdf/research/sdp/tradeinillegaltimber.pdf

5 M Colchester, M Boscolo, A Contreras-Hermosilla et al, *Justice in the Forest: Rural Livelihoods and Forest Law Enforcement*. Center for International Forestry Research, Bogor, 2004

organization that fully embraces and openly assesses SFM greatly reduces its environmental and social risk.

THE HUMAN COST OF BAD FORESTRY

The social impacts of bad forestry can be stark: cronyism, corruption and predatory business practices; poorly resourced, inflexible forestry institutions; inequitable division of forest revenues; and social upheaval caused by poor land use allocation, evictions and nomadic forest employment. The main costs for people of bad forest management are:

- *Loss of livelihood.* Forest loss and degradation affect the livelihoods of forest-dependent peoples, particularly poorer groups who depend on forests for 'social security'. Ill-considered land allocation for afforestation can also lead to a loss of resources for existing land users.
- *Loss of cultural assets and knowledge.* The culture and knowledge of many peoples, which are not always documented and which have evolved over time, diminish as forest area, access and traditional rights are reduced.
- *Rising inequality.* Increasing concentration of forest wealth in fewer hands removes the development options for many. Those who lose their forest livelihoods and become marginalized may create, and themselves suffer from, social and economic problems elsewhere, such as in cities.
- *Loss of forest asset base for national development.* Asset-stripping of forests for short-term gains wipes out any potential for forest-based strategies for sustainable development.

Failure to develop legitimate governance mechanisms and invest in social services in forested regions has, in many countries, contributed to violent conflict, illegal activities and weakening of the rule of law. Where governance mechanisms are stronger, they frequently generate negative impacts by favouring only larger enterprises while denying similar rights to poorer local inhabitants.

Efforts to improve forestry can have negative social consequences if badly handled. Environmental concerns that create sudden falls in employment from harvesting restrictions can create social problems. New laws aimed at improving working conditions can lead to pre-emptive redundancies and social problems if badly handled.

Forest managers are under increasing social pressure to avoid these negative social impacts. Business viability also crucially depends on contributing to the maintenance and improvement of good-neighbourliness, security and wellbeing. SFM is a framework of connected actions which can achieve this practically and effectively.

2.3 Playing a Part in Development

FORESTRY IS THE BEST DEVELOPMENT OPTION IN SOME RURAL AREAS

Many developing countries have land that is better suited to forestry than to farming, producing some of the world's highest tree growth rates. Much of this land may not be fully utilized at present and there may be potentials for farming/plantation mixes. In some rural areas, forestry enterprises can contribute to local livelihoods and stimulate rural economies where little other development opportunity exists.

Where forestry enterprises can raise sufficient capital, and where they can employ the best technology and management skills, they can exploit their comparative advantage to make long-term investments. Where international trading connections can be made, forestry enterprises are well placed to access markets – and notably markets which demand environmental and social benefits through forestry production – and to weather periods of low commodity prices.

Box 2.3   **What do poor people get from forests?**

- subsistence goods: such as fuelwood, medicines, wood for building, rope, bushmeat, fodder, mushrooms, honey, edible leaves, roots, fruits;
- goods for sale: all of the above goods, arts and crafts, timber and other wood products;
- indirect benefits: such as land for other uses, social and spiritual sites, environmental services, including watershed protection and biodiversity conservation.

Forest managers should be alert to opportunities to start up viable forestry enterprises in rural areas that provide local benefits, and for working to ensure good management of existing forests on lands that are better suited to forestry than agriculture or other uses.

FORESTRY CAN HELP REDUCE POVERTY

Demands on the world's forests are increasing: roughly 400 million people are estimated to rely on forests for at least part of their livelihoods (see Box 2.3).[6]

All sectors have a role to play in reducing poverty. Widely agreed objectives such as the Millennium Development Goals call for a halving of the world's population living in extreme poverty by the year 2015. Forestry has advantages over many other sectors in offering potential routes out of rural poverty and can provide resource safety nets and sometimes enterprise opportunities where little else exists.[7]

Forest resources usually contribute to people's livelihoods, rather than providing all their livelihoods needs.[8] However, many more poor forest producers, traders and workers could participate in local initiatives that offer commercial prospects. Sustainable forestry management means promoting opportunities for local enterprises where possible, increasing the contribution that forestry can make to people's livelihoods.

2.4 The Business Case for SFM

MARKET PRESSURE IS RISING

Although demand for forest products globally continues to rise, the incentives for businesses to engage in SFM have traditionally been weak. Underpriced, high-value resources, such as natural forests with good timber supplies, have often been the target for timber mining rather than management, where weak government control allows it. Stock markets still place a higher premium on companies that can secure such assets at the lowest cost, which often means those with the lowest social and environmental provisions.

However, market pressure is now one of the major influences on forest managers to improve management standards. Customers are more and more interested in knowing about the

---

6 World Bank, A *Revised Forest Strategy for the World Bank Group*. World Bank, Washington DC, 2002, http://lnweb18.worldbank.org/ESSD/essdext.nsf/14DocByUnid/403A34FDD7B9E84A85256BD00077D91B/$FILE/FSSPFinal1Nov02.pdf

7 J Mayers and S Vermeulen, *Power from the Trees: How Good Forest Governance Can Help Reduce Poverty*. World Summit on Sustainable Development Opinion. IIED, London, 2002. http://www.iied.org/pdf/wssd_26_forests_and_poverty_long.pdf

8 S Wunder, 'Poverty alleviation and tropical forests – what scope for synergies?' *World Development*, 2001, 29: 1817–1834

environmental and social impacts of forestry companies that supply them with wood and paper products. Companies which can demonstrate that they implement SFM can gain access to new markets that are closed to other companies.

CORPORATE SOCIAL RESPONSIBILITY IS GAINING GROUND

Until recently the social responsibility of a major forestry company often ended with its formal obligation to pay royalties and taxes, perhaps provide cash compensation to communities for lost assets, create a few jobs and construct some schools and health clinics. Recently, however, pressure has been applied by investment companies keen to develop a green image and wary of bad publicity and the costs of conflicts associated with environmental and social problems. Tougher legislation has also caused banks and insurance companies to consider the potential liabilities related to environmental and social performance.

There is no doubt that forest managers who pursue corporate social responsibility (CSR) through implementing SFM incur additional short-term costs. These include not only the financial costs of investment and time, but also the need for high calibre staff, who may be a critically short resource. They regard these costs as a long-term investment in the forest resource which, in time, may lead to a reduction in operating costs and improved production levels. These may occur through improved security of future production, more efficient use of machinery and workforce, and reductions in some operating costs (see Box 2.4).

Some concerns have been raised that an increased focus on corporate social responsibility favours larger companies, pushing out smaller and medium-scale enterprises.[9] The biggest corporations may find it easiest to develop, communicate and demonstrate compliance with their CSR agenda, while smaller competitors struggle even to understand the requirements. This book aims to show how SFM requirements, a key aspect of corporate social responsibility for forestry companies, can be applied at a variety of scales, not only the largest.

| Box 2.4 | **Why should a forest company practise corporate social responsibility?** |

Companies practising corporate social responsibility may see a number of benefits which ultimately affect the returns and risks for investors. These may include:

- *Lower risk* – Companies with a good environmental and social performance will be perceived as less risky by financial markets, reducing capital costs and insurance premiums.

- *Secure markets* – compliance with environmental and social standards can secure markets and occasionally secure higher prices.

- *Public reputation* – this can affect the company's social licence to operate, reducing the time required to secure government approval of, and community support for, new developments or expansion.

- *Changes in legislation* (eg tightening regulations) or changes in rules on liability for damage can imply significant costs and companies that can prepare for regulatory change will have a competitive advantage.

- *Clean technologies* are usually more efficient. Similarly, good working conditions can lead to higher productivity, fewer union disputes and make it easier to attract and retain employees.

---

9 J Mayers and S Vermeulen, *Company–community Forestry Partnerships: From Raw Deals to Mutual Gains?* IIED, London, 2002 http://www.iied.org/forestry/pubs/psf.html#9132IIED

2.5 Forestry's Future Challenges

MORE FOREST GOODS AND SERVICES FROM A MIXED LANDSCAPE

Forests and forestry are changing. Natural forests are declining worldwide, with the consequent loss of biodiversity, and global climate changes are likely to bring further drastic changes. Meanwhile, trade, technology, information systems and many human aspirations and concerns are becoming globalized. This has important implications for the future of forestry.

Demand for forest products is rising. Globally, the demand for paper products, for example, is increasing by 3–5 per cent per year. Domestic consumption is rising in many developing countries – much of it for good reasons, such as the consumption of paper for health care, education and communications.

While blocks of natural forest are decreasing, forest goods and services are increasingly supplied by a range of land use types including plantations, natural or planted mixtures and farm landscapes with trees. Meanwhile, wood fibre and similar products are increasingly produced in intensive plantations or mixed forest–farm landscapes.

MANAGEMENT FOR FOREST-BASED ENVIRONMENTAL SERVICES

Society is beginning to understand the importance of, and perhaps increase its willingness to pay for, the environmental service functions of forests. These include soil fertility enhancement and protection from erosion, the regulation of natural water supply, conservation of biodiversity, carbon sequestration and the provision of landscape amenity values.

To date, these forest environmental services have benefited local and global communities but have generally had no market. Conventionally, governments have been responsible for ensuring such 'public goods' are provided. However, government weakness and shrinking budgets constrain their ability to provide such benefits, just as public demand has risen.

Numerous examples of payments for environmental services have developed in recent years.[10] For example, payments for watershed protection services have evolved spontaneously at local levels, between forest owners and municipalities (especially in the Americas) and between forest owners and water user groups (in Asia). Various approaches have been organized at the international level to market biodiversity conservation and carbon storage.

So far the market has been mostly driven by demand for environmental services. For example, there is a growing awareness of threats to the supply of environmental services, like watershed protection. But supply side and regulatory drivers are becoming increasingly important in the developing market. Most markets for forest environmental services are in their early stages and have yet to prove robust. It is difficult to know how efficient they are as a mechanism for paying for environmental services and whether they will be sustainable in the long term. However, payments for environmental services have the potential to offer an entirely new market for the services of SFM in future.

Table 2.1 outlines some examples of the developing market for forest environmental services.

LIKELY CHALLENGES FOR SFM IN FUTURE

Sustainable forestry management in the future will have to deal with a range of challenges and alternative approaches to providing forest goods and services. These challenges are likely to include:

- The need to improve forest management practices, particularly in tropical countries. This will require a combination of better training and support to forest managers, improved incentives for

---

10 N Landell Mills and I T Porras, *Silver Bullet or Fools' Gold? A Global Review of Forest Environmental Services and their Impacts on the Poor.* IIED, London, 2002 http://www.iied.org/forestry/pubs/psf.html#9132IIED

TABLE 2.1 **Commercializing forest environmental services**

Environmental service	Mechanisms for payment	Sources of demand
Watershed protection (eg reduced flooding; increased dry season flows; reduced soil erosion; reduced downstream sedimentation, improved water quality)	Watershed management contracts; tradable water quality credits; salinization offsets; transpiration credits, conservation easements, certified fish-friendly agricultural produce	*Domestic/regional* – hydroelectric companies; municipal water boards; irrigators; water-dependent industries; domestic users
Landscape beauty (eg protection of scenic 'view-scapes' for recreation or local residents)	Eco-tourism concessions; access permits; tradable development rights; conservation easements	*Domestic/international* – local residents, tourist agencies; tourists; photographers; media; conservation groups; foreign governments
Biodiversity conservation (eg conservation of genetic, species and ecosystem diversity)	Bio-prospecting rights; biodiversity credits; biodiversity management contracts; biodiversity concessions; protected areas; development rights; conservation easements; shares in biodiversity companies; debt-for-nature swaps; land acquisition	*Domestic/international* – pharmaceutical, cosmetic and biotechnology companies; agri-business; environmental groups; foreign governments; the global community
Carbon sequestration (eg absorption and storage of carbon in forest vegetation and soils)	Carbon offsets/credits; tradable development rights; conservation easements	Domestic/international – major carbon emitters (eg electricity, transport and petro-chemical companies); environmental groups; foreign governments

implementing SFM, and development and enforcement of forest legislations which promote SFM.

- Trade-offs with food production and food security goals will need to be integrated with forestry in land use decisions. This may mean both food production in forests and planning forest/agriculture land use patterns.
- Forest products may be substituted by other materials such as metals or plastics. Ensuring that forest products remain competitive with alternative materials, while providing other goods and services demanded by SFM, will be challenging. While price and product specification have dominated consumers' choices so far, in future they may turn to wood if, for example, SFM enables forest products to demonstrate their sustainability advantages over alternative construction materials. Alternatively they may turn away from forest products on the basis of the perceived social and environmental impacts of forestry.
- Illegal logging is likely to remain a challenge to the viability of SFM, unless stamped out by international action. Trade in illegal forest products undercuts the products of SFM, making investment in good forest management less attractive.
- Recycling of forest products has taken off only in the paper industry, fuelled by legislation. Pro-recycling policies need to be much better informed about SFM so that the consumer can judge the relative merits of SFM and recycling and the ways in which they may be complementary.

- Finally, changes in technology can redefine the possibilities for all the above. For example, technological developments in products made from woodchips (plywood, veneer and MDF) have had enormous influences on the forests of south-east Asia. While woodchips permit a diversity of products, they favour uniformity in forest management. Thus, while forests and plantations may be able to provide an increasing number of products as technology develops, the implications for the type and location of forests will have to be carefully weighed up by forest managers and others with a stake in forests.

3 Standards for Sustainable Forest Management

The definition of sustainable forest management (SFM) given in Box 1.1 aids an understanding of the general idea of SFM, but it does not provide any detail about what it really means on the ground. How should different people's objectives be balanced against each other and how does the forest manager know in practice what level of 'undesirable effects on the physical and social environment' is acceptable.

Standards for SFM attempt to set out the way in which forest managers can balance these different objectives. As well as discussing types of standards and how they are developed, this chapter looks at some of the main SFM standards that are of practical interest to the forest manager, including:

- international initiatives aimed at defining SFM at the operational or forest management level;
- national initiatives that develop local interpretations of international standards.

3.1 What are SFM Standards?

Defining what constitutes sustainable forest management is complex. During the past 15 years there have been many different attempts to do so. All of these initiatives use a 'standard' made up of principles, criteria and/or indicators to define the elements of sustainable forest management (see Box 3.1).

The way in which a standard defines sustainable forest management depends on exactly what the principles, criteria and indicators require and how they are phrased. Some standards (such as the ITTO (International Tropical Timber Organization) Criteria and Indicators for Sustainable Management of Natural Tropical Forests), describe the criteria and indicators which can be used to measure trends in forest management without setting any thresholds to distinguish 'good' or 'bad' management. These may be applied at a national, regional or forest management level, providing general information about forest management in an area, without identifying whether a particular forest management unit is well managed. While these can be useful for national policy development, monitoring and reporting, they do not identify whether a particular wood product comes from a well-managed forest.

Others (such as the Forest Stewardship Council (FSC) Principles and Criteria) describe requirements with which forest managers must comply, and set thresholds for a pass/fail evaluation. The ITTO Guidelines are based around principles of good management and recommended actions for forest management, but do not stipulate concrete requirements.

Differences between standards that appear to be minor details can have a major effect on what they actually require forest managers to do. This in turn depends on who defined them and how.

3.2 Who Develops Standards and How?

Sustainable forestry management is a complex subject. It is not a precise science because we do not fully understand all the biological and physical processes of forests, or how they will respond to disturbances caused by management. In addition, SFM aims to deliver a range of objectives which sometimes conflict and need to be balanced against each other. The balance that is achieved in an

---

Box 3.1 **Principles, criteria and indicators[1]**

Principles are general in scope and outline the philosophy on which the standard is based. Examples of principles include:

Principle 1: yields of forest products shall be sustainable

Principle 2: water resources shall be maintained and conserved

Principle 3: long-term social and economic wellbeing of local communities shall be maintained or enhanced.

Criteria describe a state or situation which should be achieved to comply with the stated principle. For each principle, there will be one or more criteria. For example, to meet principle 1 (above):

Criterion 1.1: there is a continuous production of timber

Criterion 1.2: the capacity of the forest to regenerate naturally is ensured

Criterion 1.3: soil quality is maintained.

Indicators describe assessable parameters which are used to measure the state or situation expressed in the criterion. For example, for criterion 1.1 (above) the indicators are:

Indicator 1.1: balance between growth and removal of wood

Indicator 1.2: yield regulation by area and/or volume.

SFM standard will reflect the values of the people who develop the standard, and may change over time as society's values, as well as technical knowledge, change.

Finally, SFM standards that aim to be applicable worldwide need to deal with a huge variation between forest types, climate, soils, and social and economic contexts. They therefore need to be general enough to be appropriate anywhere, but with the potential to be adapted locally.

Sustainable forestry management standards cannot, therefore, be developed as a purely scientific, precise set of rules about forest management. Instead a combination of the following attributes is required:

- best available scientific information about the way forests function and respond to management;
- decision-making processes which resolve conflicting objectives of forest management and deal with gaps in knowledge about forests;
- revision over time.

Because of these factors, the people involved in developing the standard and the decision-making process used to agree the standard both have important effects on the resulting requirements. There are a number of guides to help people develop standards that are widely accepted (see Box 3.2). Some certification schemes (such as FSC and Pan-European Forest Certification (PEFC)) also have their own procedures for developing national or regional interpretations of their generic standards. These are usually based on the International Organization for Standardization (ISO) and/or International Societal and Environmental Accreditation and Labelling Alliance (ISEAL) guides.

Based on E Lammerts and E Blom, *Hierarchical Framework for the Forulation of Sustainable Forest Management Standards*. The Tropenbos Foundation, Wageningen, The Netherlands, 1997

| Box 3.2 | **ISO and ISEAL Alliance standard-setting guides** |

ISO is the International Organization for Standardization, based in Switzerland. In addition to its management system standards such as ISO 9001 and ISO 14001, ISO has produced a number of guides to setting standards and running certification schemes. ISO Guide 59: Code of Good Practice for Standardization is a widely accepted minimum requirement for developing standards in any sector. It sets out the basic requirements for standard-setting procedures, ensuring a transparent process, dealing with complaints and appeals, approval of standards, participation by stakeholders, and the advancement of international trade. However, it does not provide guidance on dealing with the complexities of forestry standards such as incomplete scientific information and resolving conflicting objectives.

The ISEAL Alliance is the International Societal and Environmental Accreditation and Labelling Alliance, and currently includes the FSC (as well as Fair Trade and Organic movements) as members. The ISEAL Code of Good Practice for Setting Social and Environmental Standards (2nd draft at October 2003) also covers good practice for standards development, specifically covering environmental and social issues. It is based on, and is compatible with, ISO Guide 59. The Code of Practice is being finalized in 2003. ISEAL members will be required to comply with the code in setting their standards; other organizations developing social and environmental standards are also encouraged to adopt it.

For more information about the development of the major standards (FSC, ITTO, PEFC, CSA (Canadian Standards Association), etc) see Nussbaum and Simula (2005).[2]

3.3 Process and Performance Standards

Two main types of standard are applied to forest management: process standards, and performance standards (see Box 3.3). Most SFM standards contain a combination of both process and performance elements.

TABLE 3.1 **Comparison of results of process and performance standards**[3]

	Process standard	Performance standard
Guaranteed minimum level of performance in the forest	No	Yes
Recognition of ongoing improvements in management	Yes	No
Management framework	Yes	No
Application to all forest types without being adapted	Yes[1]	No
Product label	No	Yes

1 In practice, the bureaucratic requirements of system standards can be a serious obstacle for small forest enterprises and for forest owners and managers who are not literate.

2 R Nussbaum and M Simula, *The Forest Certification Handbook*, 2nd edition. Earthscan, London, 2005

3 R Nussbaum, S Jennings and M Garforth, *Assessing Forest Certification Schemes: A Practical Guide*. ProForest 2002, www.proforest.net

Box 3.3   ## Process and performance standards

Process standards, also known as environmental management system standards, are based on the use of a documented management system to implement an environmental policy. No minimum level of performance is required except compliance with legislation. However, the organization makes a policy commitment to a process of continual improvement and sets itself a number of objectives and targets. Environmental management system standards include ISO 14001 and the European Union's Environmental Management and Auditing Scheme (EMAS). The Canadian Standards Association (CSA) standard for a sustainable forest management system, CAN/CSA Z809-02, contains many process elements and defines the required procedure for setting locally defined targets.

Performance standards specify fixed levels of performance that must be achieved. Forest management is evaluated against these performance requirements on a pass or fail basis. Certification standards based on the FSC's principles and criteria and those based on ITTO guidelines are examples of performance standards. However, many performance standards also contain process elements, such as requirements for planning and monitoring procedures.

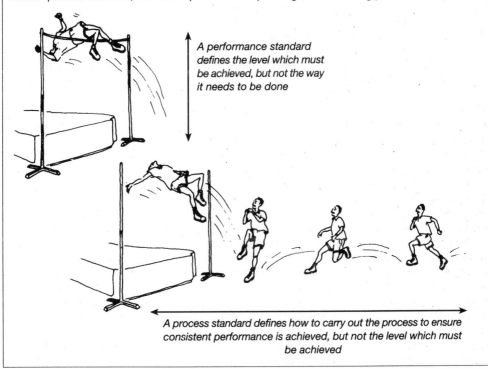

A performance standard defines the level which must be achieved, but not the way it needs to be done

A process standard defines how to carry out the process to ensure consistent performance is achieved, but not the level which must be achieved

Process standards provide a powerful tool for helping forest organizations to understand their impacts and consistently achieve their targets. They specify generic systems and are easily adapted to all forest types. However, in practice, the requirements for complex management systems, documentation and record-keeping can be an obstacle for smaller forest organizations and those run on a less formal basis. Because they specify management systems, not management results, two organizations that both meet a process standard may actually have a very different 'quality of management'.

Performance standards set out specific results that a forestry organization should achieve, without stating how it should be done. In theory, any two organizations that meet the requirements of a performance standard should be implementing the same 'quality of management'. Table 3.1 shows what process and performance standards deliver. Performance standards for forestry are subject to the limitation that it is impossible to develop a single standard that is applicable to all forests but detailed enough to be meaningful. Instead, standards must be developed regionally, or locally, within the framework of a more general international standard. This seeks to ensure that, although different, regional standards are nonetheless compatible and equivalent.

3.4 International Initiatives

There have been many international initiatives to define sustainable forest management, each with a slightly different approach. Several of these include guidelines, criteria and indicators at the level of the forest operation. These international initiatives have often been used as a basis for developing national standards or interpretations. Three major initiatives are described here:

- International Tropical Timber Organization Guidelines;
- Forest Stewardship Council Principles and Criteria;
- Pan-European Operational Level Guidelines.

INTERNATIONAL TROPICAL TIMBER ORGANIZATION (ITTO) GUIDELINES FOR TROPICAL FORESTS

ITTO is an inter-governmental organization based in Japan. It is made up of representatives of tropical timber producer and consumer countries. In July 2003, ITTO had 58 members, which together represent 95 per cent of world trade in tropical timber and 75 per cent of the world's tropical forests.[4] In 1991, the ITTO declared its Year 2000 Objective, which stated that members will progress towards achieving sustainable management of tropical forests and trade in tropical timber from sustainably managed resources by the year 2000. A review of progress towards this target, carried out in 2000, found that, although there had been considerable advances in policy and legislation in producer countries, without better field level evaluation of forest management it was not possible to say whether producer forests were being managed sustainably.[5] A new, explicit target has not yet been set.

ITTO has produced four documents concerning sustainable management of tropical forests. Each was developed by an international panel which included representatives of tropical timber consumer and producer countries, non-governmental organizations (NGOs), the United Nations Food and Agriculture Organization (FAO), consultants and academics. Because they were developed by an expert group from a wide range of backgrounds, the ITTO Guidelines have reasonably wide acceptance with government, industry and NGOs worldwide as the basis for sustainable management of tropical forests. However, some NGOs feel that the guidelines are not strong enough, particularly on environmental and social issues, and that the panels were dominated by industry and governments.

The ITTO Guidelines are relevant to operational forest management as well as to national policy, monitoring and reporting. The guidelines' general nature allows for country-specific and forest-specific interpretation throughout the tropical forest zone:

- ITTO *Guidelines for the Sustainable Management of Natural Tropical Forests* (1990) addresses natural forest management at both the national and the local forest management level. There is a strong emphasis on practical forest management for sustained yield of timber.

---

4 ITTO website, www.itto.or.jp, About the ITTO

5 D Poore and T H Chiew, 'Review of progress towards Year 2000 Objective by November 2000', www.itto.or.jp

- ITTO *Guidelines for the Establishment and Sustainable Management of Planted Tropical Forests* (1991) addresses the establishment and management of tropical plantation forests at both the national and forest management level. Biodiversity and socio-economic factors are addressed in more detail than in the natural forest guidelines.
- ITTO *Guidelines on the Conservation of Biological Diversity in Tropical Production Forests* (1993) was published after the United Nations Conference on Environment and Development (UNCED), and is consistent with both the Forest Principles and the Convention on Biological Diversity. The guidelines cover conservation of biodiversity both at the regional level and in operational production forests.
- ITTO *Criteria and Indicators for Sustainable Management of Natural Tropical Forests* (*revised* 1998) provides criteria and indicators for the assessment of changes and trends in forest conditions and management systems at the national and forest management unit levels. The original ITTO Criteria (1992) have been revised to cover the full range of economic, environmental and social goods and services.
- ITTO *Guidelines for the Restoration, Management and Rehabilitation of Degraded and Secondary Tropical Forests* (2002) aim to fill the increasingly important gap between the existing ITTO Guidelines for natural forests and planted forests. Principles and recommended actions are divided into two sections, aimed at two different levels: the policy level and the practical forest level. The guidelines emphasize rehabilitation of forests for multiple use under collaborative agreements.

The ITTO Guidelines are important because they have influenced evolving policy and legislation in tropical countries and provide a framework for countries to evaluate their forest management. They are widely accepted by governments.

THE FOREST STEWARDSHIP COUNCIL PRINCIPLES AND CRITERIA

The Forest Stewardship Council (FSC) is an international non-profit organization founded in 1993 to support environmentally appropriate, socially beneficial and economically viable management of the world's forests. The FSC is an association of members, open to anyone involved in forestry or forest products: members include representatives of environmental and social NGOs, the timber trade, forestry professionals, indigenous people's organizations and community forestry groups, as well as certification organizations. The membership is divided into three equal chambers of economic, social and environmental interests, each with equal northern and southern hemisphere representation. The FSC secretariat, which runs the FSC on a day-to-day basis, is based in Bonn, Germany. The secretariat is responsible to a board which is elected by the members.

The Forest Stewardship Council's Principles and Criteria for forest management (FSC P&C) (see Appendix 1.1) were developed by NGOs in consultation with industry and academic bodies. They were designed to provide a basis for the development of standards for use in voluntary forest certification (see Part Six). The FSC P&C focus on operational forest management; they are designed to be applicable to all forest types and to be interpreted into national or regional standards. The FSC has defined a process by which these national or regional interpretations should be made; indicators that are defined at the national or regional level are then used by all certification bodies operating in that area.

The FSC P&C consist of ten principles: nine principles are applicable to all types of forests, while Principle 10 is specific to plantation management. The emphasis is on minimizing the negative impacts of all forestry operations on the environment, maximizing social benefits and maintaining the important conservation values of the forest.

The FSC P&C are significant because they are applicable globally, to tropical and temperate forests in developed and developing countries. They are widely accepted by environmental and social NGOs, as well as many retailers of forest products.

PAN-EUROPEAN OPERATIONAL LEVEL GUIDELINES

The Pan-European Operational Level Guidelines[6] (PEOLGs) are designed to provide a forest operational level interpretation of international guidelines drawn up at inter-governmental level during the 1990s. At a Ministerial Conference in Helsinki in 1993, ministers responsible for forestry in Europe embraced the Forest Principles developed at the United Nations Conference on Environment and Development (UNCED). Two resolutions translating this into European level policy were adopted (The General Guidelines for Sustainable Management of European Forests and The General Guidelines for the Conservation of Biodiversity of European Forests).

These were later elaborated into national level criteria and indicators, which could be used for evaluating and reporting progress towards sustainable forest management at a national level. The Pan-European Operational Level Guidelines, adopted by the Third Ministerial Conference on Protection of Forests in Europe in 1998, provide a practical interpretation of the criteria and indicators, translating international governmental commitments down to the level of forest operations.

The PEOLGs are significant because they interpret inter-governmental policy initiatives at the forest operational level. The PEOLGs provide a common set of guidelines within Europe for management of forests. They have been adopted as a basis for development of national certification standards in a number of European countries.

3.5 National Initiatives

Many countries have developed their own national standards forest management. Most national standards have been based on one or more of the international standards outlined above.

The development of a local or national standard is particularly important with performance standards (such as ITTO and FSC) because no single performance standard can be applicable to all forests. Local variations and conditions must be taken into account.

ITTO-BASED NATIONAL (AND REGIONAL) INITIATIVES

Several countries have developed national interpretations based on the ITTO guidelines. The Malaysian Criteria, Indicators, Activities and Standards of Performance for Forest Management Certification (MC&I, October 2001) are based directly on the ITTO Criteria and Indicators for Sustainable Management of Natural Tropical Forests. For each ITTO Indicator considered to be relevant in the Malaysian context, the standard defines activities required to meet the indicator and standards of performance specified for Peninsular Malaysia, Sabah and Sarawak.

The African Timber Organisation (ATO), in collaboration with ITTO and the Center for International Forestry Research (CIFOR), developed a regional set of principles, criteria and indicators based on the ITTO Criteria and Indicators. The ATO/ITTO Principles, Criteria and Indicators for the Sustainable Management of African Natural Tropical Forests were published in 2003.

In Indonesia, the Lembaga Ekolabel Indonesia (LEI) developed a national standard in 1996 based on the ITTO Guidelines for natural forests.

FSC NATIONAL INITIATIVES

FSC national initiatives can develop national or regional interpretations of the FSC P&C and submit them to the FSC for formal accreditation. The national working group needs to be formally accredited by the FSC Board before a national or regional standard can be submitted to the FSC for accreditation.

---

6 Available at PEFC website www.pefc.org Reference Documents

Box 3.4	**National and regional standards accredited by FSC, March 2004**	

Country	Standard	Date accredited
Bolivia	Standards for forest management certification of Brazil nut	March 2002
Bolivia	Standard for certification of forest management of timber yielding products in the lowlands of Bolivia	September 2000
Brazil	Certification standards for forest management on 'Terra Firme' in the Brazilian Amazon	May 2002
Canada	Certification standards for best forestry practices in the Maritime Forest Region	December 1999
Canada	Regional forest certification standards for British Colombia	Preliminary standard July 2003
Colombia	Standards for voluntary forest certification of natural forests	February 2003
Colombia	Standard for forest management certification of plantations	October 2003
Germany	FSC standard	November 2001
Peru	Forest management standards for the production of Brazil nuts	October 2001
Peru	Standards for forest management certification for timber products in the Amazonian forests of Peru	May 2002
Sweden	Swedish FSC standard for forest certification	May 1998
UK	Forest management standard	October 1999
USA	Regional forest stewardship standard for the Lake States–Central Hardwoods region	August 2002
USA	Rocky Mountains regional standards	September 2001
USA	Forest certification standard for the south-eastern United States	November 2002
USA	FSC certification standard for the north-east region of the United States	November 2002
USA	Regional forest stewardship standard for the south-west region	July 2003
USA	Regional forest stewardship standard for the Pacific Coast region	July 2003

The group is required to keep a balance in its membership between the three chambers (environmental, social and economic) to ensure standards are fair. National or regional standards must fully cover all the FSC P&C and must have been developed through a consultative process.

Thirty-three national working groups had been accredited by the FSC worldwide by June 2004. Eighteen national and sub-national standards have been endorsed by the FSC, as shown in Box 3.4. In a number of countries, including Sweden and the UK, the national standard is undergoing the first revision, which is required every five years.

PEOLG-RELATED NATIONAL INITIATIVES

The Programme for the Endorsement of Forest Certification Schemes (PEFC, formerly known as the Pan-European Forest Certification Scheme) is a framework for the development and mutual recognition of national certification schemes based on internationally agreed principles of sustainable forest management. Since 2002, forest certification standards developed under the PEFC umbrella in Europe are expected to be compatible with the Pan-European Operational Level Guidelines (PEOLG).

Outside the European area, nationally developed certification standards submitted for endorsement under the PEFC scheme are required to be based on other inter-governmental criteria and indicators for sustainable forest management. These include the Criteria and Indicators for the Sustainable Management of Temperate and Boreal Forests (Montreal Process), the ITTO Guidelines for Natural Tropical Forests, the Tarapoto Agreement in South America and the African Timber Organization Criteria and Indicators for Sustainable Management of Natural Forests.

By December 2003, PEFC included 27 members on its council (each member being a governing body representing a national or sub-national scheme). Thirteen national forest certification schemes, all European, had been endorsed by PEFC.

Part Two

What Do the Standards Require?

Introduction to Part Two

Part Two looks in detail at the Forest Stewardship Council's Principles and Criteria for forest management and ITTO's Guidelines and Criteria and Indicators for natural and planted tropical forests, which were described briefly in Section 3.4. The FSC and ITTO requirements and recommended actions have been summarized and combined for simplicity: the actual standards themselves can be found in Appendix 1.

Part Two is divided into five sections which provide a brief introduction to each topic and then outline the requirements for SFM as defined by FSC and ITTO. Each section is cross-referenced to other parts of the handbook, where there is further information about implementing these requirements.

Sustainable forest management (SFM) is generally accepted as the production of a range of goods and services from the forest, without degrading the forest's ability to provide further goods and services in future. The following themes provide the structure for Part Two and can be recognized by the icons in the top right-hand corner of the page:

 Chapter 4 The Legal and Policy Framework

 Chapter 5 Sustained and Optimal Production of Forest Products

 Chapter 6 Protecting the Environment

 Chapter 7 The Wellbeing of People

 Chapter 8 Plantations

Plantations are subject to different conditions from natural forests: while the requirements outlined in Chapters 4 to 7 also apply to plantations, Chapter 8 addresses the issues specifically applying to plantations.

Using Part Two

THE REQUIREMENTS

Near the start of each section there is a table summarizing the requirements of FSC Principles and Criteria and/or the ITTO Guidelines. When these requirements overlap, they are shown on the table as *shared requirements*: if explicitly required only by one or other (FSC or ITTO), this is shown in the table.

RESPONSIBILITY

Each requirement in the table is matched to a column headed *responsibility*. This indicates the level within the forest organization's hierarchy at which responsibility for addressing this requirement is most likely to lie. Because organizations can range from very large to very small, these categories are not definitive.

P = policy level or senior management will most likely be responsible;

T = technical forest management team will probably be responsible;

S = specialist inputs may be needed.

SCALE

Forestry organizations have variable impacts, depending on many factors, such as the size of the organization, the type and intensity of forest management, the importance of the forest for biodiversity and the importance of the resource to local people for subsistence or employment. In this handbook high impact forest organizations are described as 'large-scale' and low impact organizations are called 'small-scale'.

As large-scale and small-scale organizations may not need, or be able, to implement the requirements of FSC and ITTO in the same way, the third column of each table contains one of two icons to show when scale may need to be considered:

 = all scales should consider this requirement

 = small-scale organizations may treat the requirement differently.

Where a small-scale icon appears, there is an explanation of the different ways a small-scale and large-scale organization might approach that requirement. To give an indication which scale is most appropriate, use Table Part 2-1 below.

This table is intended to assist the forest manager to think about the scale and impacts of their forest organization. The table does not provide definitive answers about the intensity of a forest organization's impacts: it is intended only as a guide to help work out when scale might need to be taken into account in meeting the requirements of SFM standards.

This table is for guidance only and does not correspond with any official FSC or ITTO definitions of scale.

HOW TO USE THIS TABLE

1 For each of the questions in the first column (headed 'identifying impacts'), work out which is the best description, by looking across its row to select the most appropriate answer. Then look at the column heading to see the score. Note this score in the final column (headed 'your score'). For example: a small forest (depending on the local context) scores 1 while a large forest scores 3.
2 Add all your scores together.
3 Compare your total score with the scale defined in the bottom row.

TABLE PART 2-1 **Determining the scale of a forest organization**

Identifying impacts	Score 1	Score 2	Score 3	Your score
1 Forest size?	small	medium	large	
2 Forest type?	managed secondary; wood lots	indigenous plantation/ old secondary	natural primary/exotic plantation	
3 Owner or concessionaire?	community	national private/state associations	international	
4 Exploitation rate?	low	medium	high	
5 Importance for biodiversity?	low	medium	high	
6 Significant environmental issues?	low	medium	high	
7 Importance for local people for subsistence?	low	medium	high	
8 Importance for local people for employment?	low	medium	high	
9 Significant social issues?	low	medium	high	
10 Past problems?	local issues only	nationally known	internationally known	
Total score				

Scale: 10–15 = small; 16–30 = medium to large

CHECKLIST

All the tables of requirements are compiled into one checklist which forms Chapter 9 of Part Two. This table has extra columns to be completed by the forest manager, recording how completely each requirement has been implemented: fully in place, partially or not at all. This will assist the forest manager to visualize how close they are to meeting the requirements.

4 The Legal and Policy Framework

Forests and forest managers are influenced by the wider political and economic climate, in particular national policy and the institutional and legal framework. Decisions on how land is allocated to forestry, or other uses, are important for the success of SFM. For example, conflicts can occur between forestry, agriculture, mining and other land uses. If the legal framework for resolving these conflicts is inadequate, the forest manager is left in a difficult situation.

Nevertheless, there are important legal and policy issues which are part of the forest manager's responsibilities including:

4.1 compliance with legislation and regulations;
4.2 tenure and use rights;
4.3 the forest organization's commitment to SFM.

4.1 Compliance with Legislation and Regulations

BACKGROUND

All forestry organizations must work within the legal and regulatory framework. The laws and regulations of the country in which they operate, as well as appropriate international laws, must be respected.

REQUIREMENTS

1 Compliance with *local and national regulations* requires compliance with all relevant legislation, such as:
   - forest legislation and mandatory codes of practice;
   - environmental protection legislation;
   - conservation and wildlife legislation;
   - land allocation legislation;
   - labour and employment legislation;
   - health and safety regulations;
   - financial and tax legislation;
   - other legislation regulating uses of the land, such as agriculture, mining, water management, recreation and access.
2 Normally, when a government signs an *international convention*, it incorporates the provisions into national legislation. However, there is often a time lapse before this happens. The forest manager should comply with the requirements of the international agreement. The main international agreements relevant to forest management are:
   - Convention on International Trade in Endangered Species – CITES (see Appendix 3.2);
   - International Labour Organization (ILO) Conventions (see Appendix 3:3);

   The International Tropical Timber Agreement (ITTA) and UN Convention on Biological Diversity are both important for forestry and are mentioned in the FSC requirements. However, there are no specific requirements of these international agreements that are relevant at the forest managers level.

TABLE 4.1 Compliance with legislation and regulations

FSC requirements: (The requirements below are not specifically required by ITTO, but they are implicit in ITTO's guidelines.)	Responsibility	Scale
1 Compliance with local and national regulations	P	
2 Compliance with applicable international agreements	P	
3 Payment of all charges, fees and royalties	P	

3 *Payment of all applicable charges* such as fees, royalties and taxes. A forest organization which evades paying such charges cannot be considered to be contributing to national development or responsible forest management.

Compliance with legislation and regulations – see also:

Section 7.2: Social Impact Assessment

Section 15.1: Environmental and Social Impact Assessment

Appendix 3: International Conventions

4.2 Tenure and Use Rights

BACKGROUND

Sustainable forest management is a long-term undertaking, requiring investment and protection of the forest resource throughout the operation. Long-term, legal tenure or use rights to the forest area is therefore essential, although there are often several different claims to rights over forested lands.

A forestry organization must ensure that both the legal and traditional use rights of local communities are recognized and respected. Even where these rights are not reflected in formal laws, they are still significant for SFM.

TABLE 4.2 Tenure and use rights

Shared FSC/ITTO requirements:	Responsibility	Scale
1. Long-term legal rights to manage the forest resource	P	
2. Recognize and respect local communities' legal or customary rights	P	

REQUIREMENTS

1   The forest organization must have a long-term legal right to manage the forest area. Where land is held on a short-term lease, the forest organization should work with the owner to secure long-term rights or ensure long-term management.

   Land tenure and use rights may be demonstrated by:
   - ownership documents;
   - concession agreement;
   - cutting licence;
   - proof of customary rights.

2   Forest management operations, especially timber harvesting, can have a detrimental effect on other users of the forest, especially indigenous and local communities, who often rely on the forest resource for subsistence, or other goods and services. If other people have legal or customary rights to the forest resource, their free and informed consent to the forest operations must be obtained before operations begin.

Tenure and use rights – see also:

Section 7.3: Recognition of Rights and Culture

Section 15.2: Communication and Collaboration with Stakeholders

Section 15.3: Writing a Management Plan

Section 19.5: Agreeing Social Responsibilities

4.3 The Forest Organization's Commitment to SFM

BACKGROUND

Because SFM is a long-term process, some of the financial returns of forestry production must be reinvested. Commitment is needed to ensure that adequate investment is made to help develop the management capability of staff, maintain the ecological productivity of the forest and support sustainable development of local communities.

REQUIREMENTS

1   SFM must be regarded as a long-term process which requires constant *reinvestment of financial benefits* from production into the maintenance of the system. This may involve investment in training for staff, developing management systems and research activities, continued stakeholder participation, or direct investment in silvicultural treatments of the forest.

   It does not mean providing limitless money and people – the resources available will depend on the size and wealth of the forest organization. Appropriate investment frequently does not involve capital expenditure on high profile infrastructure: investment in training and development of people may be more effective. Inputs should be planned and funds allocated in advance to match income generation.

TABLE 4.3 Commitment to SFM

Shared FSC/ITTO requirements:	Responsibility	Scale
1 Reinvest part of the financial benefits from forest management in maintaining SFM	P	🌲
FSC requirements: 2 Demonstrate long-term commitment to SFM	P	🌲

2   FSC requires that a *commitment to* SFM *must be demonstrated* at all levels of the organization. In large organizations, the technical forest manager is often familiar with the requirements of SFM but other levels within the organization hierarchy may not fully understand what is needed. When this is the case, forest managers should spend time ensuring that each management level understands and is committed to SFM. For example:

- Commitment is particularly important at the highest levels, as it is chief executives, directors, owners and shareholders who make the strategic decisions affecting the availability of funds and staff for SFM.
- It is also essential that the aim of SFM is understood and supported at the operational level by chainsaw operators, field supervisors, and those in charge of community relations, as their activities directly affect the way the forest is treated.

Commitment is usually initially demonstrated in the form of a policy statement, which outlines the broad principles of the operation. This can be used to communicate SFM commitment both within and outside the forest organization. Ultimately, however, allocation of resources to activities required for SFM will demonstrate the forest organization's commitment.

Commitment to SFM – see also:

Section 10.1: Developing Commitment

Section 10.2: Writing a Policy Statement

Section 15.2: Communication and Collaboration with Stakeholders

5 Sustained and Optimal Production of Forest Products

SFM means more than ensuring the sustained production of timber or non-timber products (NTFPs) from the forest. Overall planning of forest operations, definition of sustainable yields to be harvested, monitoring the effects of management, protection of the forest from illegal activities and optimization of the range of benefits derived from the forest are all essential. This chapter focuses on:

- management planning;
- sustained yield of forest products;
- monitoring;
- protection of the forest;
- optimizing benefits from the forest.

5.1 Management Planning

BACKGROUND

Sustainable forest management involves the achievement of multiple management objectives. These objectives and means of achieving them should be defined in a forest management plan or equivalent documents.

1 *Management planning* is the foundation of a well-managed forest. It lays out what is to be done, where, when, why and by whom. The management plan should be authorized by senior management (eg the managing director) of the forest organization. Management plans should be working documents and should be available to and used by everyone who has a decision-making role. An example of the contents for a forest management plan is given in Appendix 2.1.

A forest management plan usually applies to a particular forest management unit (FMU). An FMU is an area of forest under a single or common system of management, which is described in the management plan. The FMU might be a large contiguous forest concession or a group of small forestry operations, possibly with different ownership. The important element is the common system of management.

TABLE 5.1 **Management planning**

Shared FSC/ITTO requirements:	Responsibility	Scale
1 Undertake management planning at appropriate levels	T	🌱
FSC requirements:		
2 Periodically revise the management plan	T	
3 Make a summary of the management plan publicly available	P	🌲

Depending on the size of the operations, management planning is often broken down into three levels:

- Strategic plan: The forest management plan for an entire forest operation, over the long-term, such as an entire rotation, or a 25-year period.
- Tactical plan: Sets out the activities planned, normally over a five-year period, in more detail. It may coincide with business plans.
- Operational plan (Annual Plan of Operations): Details the precise activities to be carried out over the next year. This will include month-to-month activities and should provide the most direct control over operations: it is fundamentally important in ensuring efficient and environmentally acceptable operations. It may coincide with annual budgeting.

The tactical and operational plans must be compatible with the strategic plan; they are derived from it, but provide more detailed prescriptions. In some countries, the strategic and tactical plans are combined into one document, sometimes known as a 'medium-term' plan which covers a period of 10–20 years. Local regulations may dictate the content and structure required for the management plan (see Box 5.1).

Scale considerations!

The detail and length of the management plan should be appropriate to the scale of the operations. Producing three separate plans may be inappropriate for a small, low impact operation – instead they could be combined as one document.

2 Once written, a *management plan should be revised* periodically to ensure that it does not become a static and unchangeable document. Revisions should take account of changing situations, new information and technology. To do this, there must be a mechanism for channelling information from the monitoring programme into management plan revision. Each revision is an opportunity for the forest manager to revise objectives and methods. The Environmental Management System

Information for revising the management plan may come from a variety of sources

(EMS) approach described in Part Three should help forest managers to develop a dynamic and responsive approach to management planning.

3 FSC-based standards require that, while respecting the confidentiality of certain commercial information, *a summary of the management plan should be available for the public*. This improves the transparency and accountability of forest operations and promotes better communication with stakeholders. The summary must be appropriate to the readership: technical language should generally be avoided while illustrations may be usefully added.

Management planning – see also:

Section 15.3: Writing a Management Plan

5.2 Sustained Yield of Forest Products

BACKGROUND

A prerequisite of sustainable forest management is that removal of forest products does not exceed levels of regrowth. In commercial forests where the major product is timber, this means calculating and implementing sustained yields for timber harvests. This requires information which shows stocking levels and replacement rates (for example, inventory data and growth and yield data) and which can be used as a basis for calculating sustainable harvest levels.

Although this handbook focuses on timber production, in areas where NTFPs are harvested, similar inventory data and calculations will be needed to ensure that harvesting levels remain within the capacity of the forest for replacement. In many countries methods for calculating NTFP yields are not fully developed and research support may be needed.

REQUIREMENTS

1 Harvest rates must be set at *sustainable levels*. This is the fundamental basis of SFM. Where possible, harvest levels should be set for each main wood and non-wood forest product. However, determining the sustainable harvest level for natural forests is complex, because this level is likely to change as the forest grows and is harvested, and as markets for different species change.

In natural forests, the first harvest consists of accumulated timber 'capital' which has built up over a long period. After harvesting there is a major change to the species composition and size class mixes. More explanation about why harvest levels should change over time is given in Box 5.1

2 *Reliable data* must be collected in order to determine a sustainable harvest level. Documents and records must be maintained to assist monitoring and to determine trends over time. This information needs to be analysed, and to hand, in order for it to be useful. Key data include:
   - forest inventory data, providing information about the quantity of harvestable resource currently available;
   - information about growth and yield, to determine how fast the resource will be replaced after harvesting;
   - information about seed production and regeneration from ecological studies and background knowledge;
   - information about NTFPs if these will be harvested or may be affected by harvesting operations;
   - information about non-timber species used by local people or harvested by the forest organization;
   - information about conservation values including High Conservation Values that may be present and affect management, and potential keystone species (see Box 6.1)

TABLE 5.2 **Sustained yield of forest products**

Shared FSC/ITTO requirements:	Responsibility	Scale
1 Set harvest rates at sustainable levels	T	
2 Collect data defining sustainable production levels	S	
3 Adopt a reliable method of controlling yield (eg AAC). Where data are unreliable, set production levels conservatively	T	
4 Maintain records of actual production levels of wood and non-wood products	T	
5 Periodically revise yield levels	S	
6 Document and justify the choice of silvicultural system	T	
7 Properly supervise all harvesting operations and silvicultural prescriptions	T	

Scale considerations!

It is not always possible for a forestry organization, especially smaller operations, to carry out all the necessary research for defining a sustainable harvest level. Growth and yield predictions require careful design of permanent sample plots; there must be precise measurement and remeasurement of plots; and sophisticated analysis and modelling to provide reliable predictions of future growth and yield. In many countries the government forestry department or research institutes bear responsibility for gathering this type of data. The forest manager should liaise with the institutions responsible. Collaborative research may be the most cost-effective means of obtaining data.

Large companies should consider specialist inputs to data collection and analysis, developing their own research capacity and collaborative relationships with research institutions.

Small forestry organizations should, at a minimum, carry out an inventory of the forest resources available (timber and/or non-timber, depending on expected production). Growth and yield predictions and information from existing ecological studies should be compiled, using other relevant material developed in the region.

Box 5.1 **Sustainable harvest levels in natural forest**

When primary forest is first logged it normally contains a high standing volume of timber, a variable proportion of which is marketable, depending on composition and market demand. Because this standing volume has accumulated over a long period, the commercial timber is likely to be of a quality and volume that will probably not be matched in future cuts (because it contains slow-growing specimens and species, large diameters, and so on) unless the logged forest is closed to further exploitation for a century or more. In this sense the first crop is, in practical terms, not repeatable.

If production of timber is to be genuinely sustainable, the single most important condition to be met is that nothing should be done that will *irreversibly reduce the potential of the forest to produce marketable timber* – that is, there should be no irreversible loss of soil, soil fertility or genetic potential in the marketable species. It does not necessarily mean that no more timber should be removed in a period of years than is produced by new growth; over-cutting in one cycle can, at least in theory, be compensated by under-cutting in the next or by prolonging the cutting cycle.[1]

---

1 D Poore, *No Timber Without Trees; Sustainability in the Tropical Forest*, Earthscan, London, 1989

3 There must be a *reliable method and sound rationale* for the calculation of the rate of harvest. A commonly used calculation, particularly for timber harvests in natural forest, is the annual allowable cut (AAC). The AAC is the volume of timber which may be cut in one year in a given area. Calculation of the AAC is based on the volume of timber in the area which can be harvested, while leaving enough stems to provide the next crop. It depends on the standing stock, the growth rate and the size of the forest operation. The harvest level must be set conservatively where growth and yield data are unreliable. Some software programs, such as MYRLIN (see Appendix 4.1), have been developed to assist in calculating the AAC with limited information about tree growth rates.

   Division of the FMU into blocks or compartments and definition of annual cutting areas and volumes are essential for the practical control of the harvest level. Once the AAC has been achieved from a particular area, the blocks must be closed off and no more harvesting carried out until the next felling cycle, calculated according to the growth rate and standing stock. Premature re-entry of harvested blocks should not be permitted.

Scale considerations! ⚲

The AAC calculation is useful for larger forest organizations, because it assumes that a certain proportion of the forest area will be harvested annually. Blocks are closed after harvesting to allow regeneration. Harvesting operations rotate around the forest management unit and return to the original blocks once sufficient regrowth has occurred. In small forest organizations this may not be possible. A manager of a small area of forest may prefer to harvest the entire area at one time and then leave the area to regenerate for a given time. This may be acceptable for small areas; however, adequate provision must be made for the protection of the forest during the intervening period. Areas of forest which have been logged, and appear to be abandoned, are often encroached upon and damaged by 'salvage' loggers and small farmers.

4 *Records of production levels* of wood and non-wood products must be maintained for each compartment or block harvested. Records should be reconciled against predicted yields in order

to ensure that the AAC is not being exceeded. This information is also essential for the prediction of future growth and yield and the accurate revision of yield levels.

The maintenance of records of the extent and nature of forest management activities greatly helps provide continuity of management when individual forest managers change.

5 *The yield must be periodically revised*, because:
  - as accumulated 'capital' is removed, especially after the first cut in natural forests, the permitted harvest level will usually change (see Box 5.1);
  - changed conditions in the forest after harvesting will affect growth rates (monitoring is required to determine these effects);
  - increasing knowledge and better modelling of growth and yield will improve the accuracy of yield estimates;
  - changing markets and technology will affect the species and log sizes that can be utilized;
  - chance events, such as fire or storm damage, may affect production targets and productive area;
  - circumstances outside the control of the forest organization may change the area available for sustained yield production (eg excisions of areas for agriculture or other land uses).

6 *The silvicultural system must be documented and justified.* This should be included in the forest management plan. The process of justifying the choice of silvicultural system helps to ensure that all levels of management understand what they are doing and why. The description of the silvicultural system should also refer to the harvesting procedures to be used.

7 Harvesting operations and silvicultural guidelines must be *properly executed and supervised.* Although a forest organization may have adequately documented harvesting procedures, silvicultural prescriptions and an accurately calculated AAC, they may not be implemented in practice. There is a high risk of this when documentation is prepared by an external consultant or government officer. It is therefore important to ensure that external specialists work closely with forest staff, to ensure that any recommendations are practical, relevant and will be implemented.

Where harvesting operations are properly supervised and monitored, the need for extensive – and expensive – post-logging assessments may be reduced (see Section 5.3).

Sustained yield of forest products – see also:

Section 15.4: Resource Surveys

Section 15.5: Calculating Sustained Yield

Section 16.3: Silviculture

5.3 Monitoring

BACKGROUND

Monitoring is essential to inform the forest manager about the effects of management within the forest management unit and surrounding area. It is particularly important to relate monitoring to the management objectives: are they being achieved? The results of monitoring should feed back into the planning process as forest management cannot be improved without data on its impacts.

TABLE 5.3 **Monitoring**

Shared FSC/ITTO requirements:	Responsibility	Scale
1 Monitor environmental, financial and social effects of operations	T/S	
2 Carry out post-harvest assessments	T	
FSC requirements:		
3 Make a summary of monitoring information publicly available	P	

1 The forestry organization should regularly *monitor the environmental, financial and social aspects of its operations.* Monitoring should be addressed at two levels: operational and strategic monitoring (see Chapter 17).

An operational monitoring programme should provide information on whether appropriate procedures are being followed and management objectives are being met. Operational monitoring should reveal management activities where there are problems or successes that could be built on. The results of monitoring must be fed back into planning.

Strategic monitoring should provide data about the long-term effects of the forestry operation, so that potential problems can be rapidly identified and resolved. Long-term monitoring of yields, growth rates and regeneration is especially important to ensure that harvest levels and mixes of products are sustainable.

To provide useful information, the records and data produced by monitoring need to be comparable over time. Monitoring procedures should be developed which are consistent and replicable. Records must be maintained and results analysed.

The following aspects are generally the most important components of a monitoring programme:
- yields of all harvested products;
- growth rates and regeneration;
- impacts on flora and fauna in production forest;
- environmental and social impacts of operations;
- effects on the quality and quantity of water derived from production areas;
- costs, productivity and efficiency;
- activities of contractors.

Scale considerations!

How much monitoring a company should carry out depends on the type and scale of operations. Some aspects, such as costs, productivity and activities of contractors (ie operational aspects) should be monitored as a normal part of management. However, some monitoring requires specialist inputs: particularly measurements of growth rates and regeneration as well as environmental and social impacts.

Large companies must devise their own monitoring programmes, employing specialist skills where necessary. Social impacts need some monitoring to be carried out independently of the forest organization. Long-term, ecological research requires efficient systems for data management and analysis. Provision must be made for the long-term funding of monitoring systems.

Small to medium-scale companies cannot be expected to carry out or fund extensive monitoring programmes independently and should consider collaborative relationships with research institutions.

2 *Post-harvest assessments* are a particular type of operational monitoring. Because harvesting and extraction operations have the greatest potential for environmental damage of any forest operation, post-harvest assessments are emphasized (see also Section 17.1).

Post-harvest assessments may evaluate:
- implementation of harvesting guidelines and criteria;
- environmental effects of harvesting operations;
- regeneration of harvested species;
- quantity of abandoned logs and waste.

Post-harvest assessments should be undertaken on a sampling basis: a selection of harvested sites should be checked. This is usually done about a year after harvesting to allow time for mortalilty caused by the harvesting to become evident. A systematic method or procedure for assessment needs to be developed. The results of post-harvest assessments should feed back into improved harvesting practices and planning of silvicultural interventions. The results might also be used to determine incentives to reward careful operators.

Good supervision during harvesting operations, combined with clear guidelines for operators, will reduce the detrimental effects of harvesting and abandonment of logs. This will greatly reduce the need for intensive post-harvest assessment.

3 FSC-based standards require that, within normal limits of confidentiality, a *summary of the results of monitoring must be made publicly available.* This permits informed discussion about the effects of the forestry operation and assists in consultation and collaboration processes (see Section 7.1).

Monitoring – see also:

Chapter 13: Checking

Chapter 17: Monitoring

Chapter 21: Monitoring Social Impacts

5.4 Protection of the Forest Resource

BACKGROUND

Sustainable forest management is a long-term activity which requires constant protection of the forest resource. Conflicts with other uses and illegal activities are major reasons for past failings of SFM. The forest needs to be protected from destructive and illegal activities which are incompatible with the objectives of SFM.

REQUIREMENTS

1 The forest resource must be *protected from illegal harvesting* and other activities which are incompatible with SFM. The forest manager's ability to do this may, however, be dependent on the legal or customary rights situation (see Box 5.2).

Forest access roads may need to be protected from use by people whose presence is not compatible with SFM, for instance illegal logging operations, settlers, illegal farmers, hunters or miners. Increasing access also increases the danger of fire and, in plantations, problems can arise from illegal harvesting, grazing and firewood gathering.

TABLE 5.4 Protection of the forest resource

Shared FSC/ITTO requirements	Responsibility	Scale
1 Protect forest from illegal harvesting, encroachment and activities that are incompatible with SFM	T	
2 Control inappropriate hunting, fishing, trapping and collecting	T/P	
3 Establish a fire management plan and warning systems for the forest organization	T	

Where encroachment is likely to pose a serious threat to long-term sustainability, a formal plan may need to be made by legal users to protect the forest. However, it is also essential that protection of the forest from illegal activities should be considered hand in hand with the protection of other users' legal or customary rights. Heavy-handed 'protection' from other people's activities may cause more problems than it solves.

Often the best means of resolving these situations is through involvement of local communities, combined with control of road access to the forest. Some of the following measures may be effective:

- involving local communities in order to reach a compromise over control of activities in the forest;
- encouraging local communities to implement their own controls to manage community use rights;
- putting up signs on access roads explaining under what conditions access is permitted;
- closing unused roads, where legally permitted, through means such as removal of bridges and culverts;
- placing guards at all entrances to the forest area.

2 Inappropriate or illegal hunting, fishing, trapping and collecting must be controlled. Where control of such activities is the government's responsibility, the forest manager should develop collaboration with the authorities, alert them to transgressions and facilitate access by enforcement personnel.

Where the forest manager does have authority to control other activities, three groups should be considered:

(i) *Communities living within the forest management unit*, which may have legal or customary rights to hunt, fish and collect. The rights of local communities must be respected and traditional hunting or fishing rights should be preserved. Community measures for controlling harvest levels should be supported. Communication and collaboration should be used to ensure hunting is legal and appropriate.

(ii) *External communities*, which use access routes provided by forest harvesting activities to enter the forest operation to hunt, fish or collect, should be monitored and their activities controlled if possible, as this can be a serious threat to biodiversity. The forest manager should enforce, or assist enforcement of any laws controlling collecting.

(iii) *Forest workers* should generally be discouraged or prevented from hunting. Wildlife populations are often severely reduced, and non-target species disturbed, around logging camps if the staff are allowed to hunt. Alternative sources of protein must be provided.

3 In fire-prone regions, *protection from fire* is extremely important. Harvesting opens up the canopy, changing the microclimate, and brash and dead vegetation may be left on the site. In these circumstances, even in relatively humid zones, the risk of fire is greatly increased. Plantations are particularly vulnerable to fire.

| Box 5.2 | **Addressing common problems with illegal activities** |

Protecting the forest resource from illegal activities can be a difficult task for forest managers. This is because:

- The forest manager may be responsible for managing the timber resource sustainably, but may have no jurisdiction over other users of the forest. For example, in British Columbia, Canada, the forest manager is responsible for managing timber resources but has no control over wildlife management or hunting.

- Legally defined rights of local people may conflict with their customary rights: what may be considered illegal by the law may be legal within an indigenous framework. For example, indigenous people may utilize areas outside their 'legally' demarcated lands for hunting, fishing or collecting other forest products, although national laws may not recognize this.

- Legal land uses may be incompatible. This is particularly common where land allocation laws are inconsistent. For example, in some countries both forestry and mining leases can be issued on the same piece of land, although the two uses are incompatible.

In such situations, the forest manager can only strive to find a working balance between the needs or demands of other users, and the requirements for protection of the forest for SFM. The important point is to consult forest users, try to reach agreement and to record the reasons for the approach which is taken.

Where fire may be a hazard, the forest manager must ensure that a practical fire management plan exists and is implemented. Adequate fire-fighting equipment and trained staff must be available. Regular tests of fire-readiness should also be carried out to ensure fire protection capability is adequate. Other forest users and local communities must be aware of fire risks, avoidance techniques or warning measures. (For more information, see ITTO Guidelines on Fire Management in Tropical Forests, 1997).

Protection of the forest resource – see also:

 Section 15.2: Communication and Collaboration with Stakeholders

 Section 19.5: Agreeing Social Responsibilities

 Chapter 19: Working with Stakeholders

5.5 Economic Viability and Optimizing Benefits from the Forest

BACKGROUND

Forest management frequently aims to maximize production of timber, even though it is just one of a multitude of products and services which different stakeholders may seek from the forest.

A central aspect of SFM is the need to optimize the mix of products and services from the forest.

Optimizing benefits in the long term involves making trade-offs between benefits which can be reaped today, and those which should be left for the future. It also requires a compromise between saleable products, such as timber, fruits and NTFPs, and vital services, such as watershed protection and wildlife habitat. These services often do not have a market and therefore the forest organization is not financially rewarded for their maintenance. However, in a number of countries, innovative markets are being developed for environmental services.

REQUIREMENTS

1 Forest management must be *economically viable*. This is distinct from the financial viability of forest exploitation. Full economic viability must take account of the reinvestment required for maintenance of the system and the additional costs (or income forgone) due to protection of the forest ecosystem and equitable distribution of social costs and benefits of forest management.

TABLE 5.5 Economic viability and optimizing benefits from the forest

Shared FSC/ITTO requirements:	Responsibility	Scale
1 Ensure forest management is economically viable, taking into account full environmental and social costs	T	
2 Safeguard multiple benefits of forests during all operations	T	
FSC requirements:		
3 Encourage optimal use	P	
4 Encourage local processing	P	

Box 5.3	**Market and non-market benefits of forests**

Forests provide a wide variety of goods and services. Some of these goods (for example, timber, nuts or honey) can be sold on the open market and therefore have a known price attached to them. Other goods may be collected and used in the household, but are not sold (for example, firewood). Such goods still have a value and a market price can usually be calculated for them by considering how much it would cost to buy them if they could not be gathered for free.

Forest services include other benefits provided by forests such as watershed protection, wildlife habitat and landscape values. They are more difficult to value and are rarely bought and sold on the market. In addition, the costs of conserving forest services often fall to a different group of people from the benefits of using them. For example, the costs of protecting a watershed may fall on the forest manager in terms of restricting logging on steep slopes to prevent erosion, or construction of bridges and culverts to high standards to prevent siltation or damage to watercourses. The benefits of watershed protection may be felt by other people external to the FMU – in this case, clean water or fisheries for communities downstream.

In a few cases, governments have put a value on the external non-market benefits of forests. In Costa Rica, for example, since 1996 the government has made payments to landowners for the environmental services which their forests provide to society, based on the services of erosion prevention, carbon storage, and watershed and biodiversity protection. Average payments are approximately US$50/ha/year.[2]

SFM is about managing for a mixture of goods and services and optimizing the balance between them, bearing in mind that financial returns and market prices are not the only indicators of value.

Implementation of SFM is rarely as financially attractive in the short-term as unsustainable forest exploitation. Forest exploitation pays no regard to the non-financial services provided by forests, such as protection of soil and water resources and provision of unmarketed services to local communities (see Box 5.3). In the long term, however, provision of these services is considered to be an essential prerequisite of SFM. In some situations, markets are developing for environmental services, which may help to pay for some of the costs (see Chapter 2).

2 Forestry operations must ensure that the *multiple benefits of forests are safeguarded*. These benefits include:

- *Soil and water resources*: streams and rivers are often important resources for people living downstream. 'Non-market benefits' such as clean water, or fisheries spawning sites, are resources that must be protected during all forest operations.
- *Biodiversity values*: forests often provide important wildlife habitats and can protect an important range of biological diversity (see Section 6.2). This is especially true in natural tropical forests, which are particularly rich in plant and animal species.
- *Recreation and education*: forestry organizations can promote recreational and educational use of the forest. This can be a useful way of involving local people and building a positive relationship between the organization and local communities.
- *Food, fibre and fuel*: the forest may be an important source of food, fibre and fuel for local communities, and for this reason they should continue to have access to the forest for traditional activities. Hunting, fishing, trapping and collecting should be at a scale appropriate to domestic consumption. Where activities exceed subsistence scale, inventories and agreed harvest levels are needed to prevent over-harvesting.

---

2 V Watson et al, *Making Space for Better Forestry: Policy That Works for Forests and People*. IIED/Centro Cientifico Tropical, San José and London, 1998

Box 5.4	**Cost-benefit analysis of options for forest management**

Cost-benefit analysis (CBA) is a method used by economists to try and compare the achievement of objectives with the costs incurred.[3] It attempts to value market and non-market benefits (see Box 5.3) in a common currency (usually monetary) and compare the costs and benefits of managing for different mixtures of objectives over time.

As a simple example, it might mean weighing up the long-term costs and benefits of improved road maintenance:

- on one hand the present additional costs of increased labour and machinery to improve road maintenance now;

- on the other hand the future costs of increased downtime and higher repair costs to trucks; income forgone due to slower transportation of products; and non-market costs of increased siltation of watercourses from run-off from poorly maintained roads.

CBA can be a complex process itself, but thinking in terms of costs and benefits – and considering both marketable and non-market goods and services of forests – can be extremely effective in helping the forest manager think clearly about where their costs and benefits lie.

- *Inter-cropping, mixed cropping and livestock grazing*: managers might consider allowing these activities by local people on suitable areas. The costs and benefits of such mixed systems need careful evaluation (see Box 5.4), however, as the growth of crops and/or trees may be affected by mixing livestock or crops with trees.
3 FSC-based standards require that *optimal use* of the forest's multiple products is encouraged by forest management. Over-reliance on a single or a few timber species can lead to inefficient harvesting and 'creaming' of valuable species leading to genetic degradation, as well as an insecure market, vulnerable to changes in prices for particular species.

3 R T Prinsley, *Agroforestry for Sustainable Production: Economic Implications*. Commonwealth Science Council, 1990. ISBN 0-85092-342-5. Available from the Commonwealth Secretariat Publications.

Box 5.5 **Addressing common problems with lesser-known species**

Production of timber from tropical natural forests is frequently based on relatively few species. For example in Subri Forest Inventory, Ghana (described in Plumptre, 1996),[4] 135 species were listed as producing logs; of these 35 occurred in potentially commercial volumes, but only 10 species were actually commonly used. Similar situations exist in other regions.

While harvesting a wider range of species might be preferable in order to prevent over-reliance and over-harvesting of particular species, the following points need to be considered:

- the actual volumes (total per hectare and of individual species) which can be sustainably harvested need to be determined;

- the felling cycles appropriate to harvesting lesser-known species;

- the markets, local and export, which will accept a wider variety of species.

However, a wider range of species may be difficult to market and attention needs to be given to the possible effects of higher per hectare volumes harvested if a variety of species is utilized (see Box 5.5). Marketing efforts need to assist the forest manager to harvest and sell a wide range of products, with consideration given to sustainable production of both timber and non-timber products.

4. FSC-based standards require that *local processing* is encouraged, although the definition of local remains vague. This is intended to increase opportunities for local people, living within daily travelling distance of the processing facility and neighbouring communities, to benefit from forest

---

Box 5.6 **Local processing: some pros and cons**

Small-scale, local processing and cottage industries located near sawmills have the advantage of:

- providing jobs, skills and opportunities for local communities;

- ensuring that a proportion of the benefits from the forest resource is received by local communities;

- providing an opportunity for the community to add value locally.

However, the ability to add economic value to wood products depends on the local situation. Processing facilities in remote locations face particular problems including:

- finding spare parts, new machinery and skilled labour; this may result in poor quality processing and in value-lost rather than value-added;

- poor quality sawmilling and processing can cause excessive waste of the resource;

- local communities may become over-dependent on the processing facilities and therefore on the forest operation. Harvest rates may be set by the needs of the processing centre and not the capacity of the forest.

Where adequate roads and transport are available, log transport to nearby towns may become more attractive than remote, on-site processing facilities. This also has the benefits of better provision of services such as electricity, water, medical facilities and housing as well as proximity to markets and labour.

---

4 R A Plumptre, Links Between Utilisation, Product Marketing and Forest Management in Tropical Moist Forest. *Commonwealth Forestry Review*, 1996, 75: 316–324

management through increased employment and income generation from processing. If successful it can act as a nucleus for the development of supporting businesses, promoting further development. However, the practicalities of local processing must also be considered (see Box 5.6).

Optimizing benefits – see also:

 Section 15.2: Communication and Collaboration with Stakeholders

 Section 15.5: Calculating Sustained Yield

 Section 19.5: Agreeing Social Responsibilities

6 Protecting the Environment

Forestry can have a major impact on the environment, both within and outside the forest management unit. Where possible, the environmental benefits of forests should be maximized and adverse impacts minimized.

Protecting the forest includes conserving biodiversity, maintaining its ecological functions, protecting soil and water resources, and minimizing waste and pollution. In many countries there are local guidelines on environmental protection; if these are available they should be adopted as a minimum requirement. This chapter considers the need for:

- environmental impact assessment;
- conservation of biodiversity;
- ecological sustainability;
- use of chemicals;
- waste management.

6.1 Environmental and Social Impact Assessment

BACKGROUND

An environmental impact assessment (EIA) aims to look at forest management and:

- identify actual and potential impacts on the environment as a result of operations;
- plan to minimize or avoid negative environmental impacts and maximize positive environmental impacts.

Similarly a *social impact assessment* (see Section 7.2) aims to look at the forest organization's plans and:

- identify actual and potential impacts on people as a result of operations;
- plan to minimize or avoid negative impacts and maximize positive impacts on stakeholders and others.

Because environmental and social impacts often overlap, a social impact assessment is often carried out at the same time as the EIA. This process is referred to as an environmental and social impact assessment (ESIA).

TABLE 6.1 Environmental and social impact assessment

Shared FSC/ITTO requirements:	Responsibility	Scale
1 An ESIA should be carried out prior to site-disturbing operations	T/S	🕯
2 ESIA results must be integrated into management operations	T	🏕

REQUIREMENTS

1 An *environmental and social impact assessment* should be carried out before any site-disturbing operations are carried out. Where operations are already on-going, an ESIA should be carried out for all existing and future operations. In many countries an EIA or ESIA is a legal requirement. If local guidelines already exist, they should be followed.

Environmental and social assessments may be appropriate at a variety of scales: for example, a full ESIA might be carried out for an entire forestry organization as part of its initial planning process. A limited environmental assessment may be appropriate to individual operations, and might be carried out on a routine basis. For example, completion of an environmental assessment prior to establishment of a new extraction track or minor road is a means of identifying any potential problems and ensuring that environmental planning guidelines are being followed. (See Section 15.1 for more details on both sorts of ESIA.)

An ESIA can be useful if it is prepared by the forestry organization and used as a way to start a programme of building environmental awareness and commitment internally. However, an ESIA carried out by an external organization may carry more credibility outside the forestry organization, as third party approval and monitoring are often important. In countries where an ESIA is a legal requirement, it is usually carried out by an external organization.

An ESIA usually includes the following steps:

- setting the scope of the ESIA, to identify the aspects of forest operations that are likely to give rise to key issues;
- collecting baseline data on the aspects identified during scoping, to provide a basis for the evaluation of potential or actual impacts, and to act as a baseline against which to monitor subsequent changes;
- identifying impacts and evaluating their significance, in order to decide which areas should be priorities for immediate action;
- identifying mitigating or alternative actions, to reduce or avoid negative impacts of forest operations and setting objectives and targets for implementing them.

Stakeholder consultation is essential throughout the ESIA process, from identifying the scope of the ESIA, through to discussions of options and alternative actions.

Scale considerations! ⚲

Medium to large-scale operations should undergo an independent ESIA by an external organization able to provide a multi-disciplinary team to consider fully all aspects of operations. Small-scale, low impact forest organizations may consider carrying out and documenting their own ESIA. This may be less formal than an independent ESIA, but should aim to cover the main steps outlined in Section 15.1.

2 The results of the ESIA *should be clearly documented and integrated* into forest management planning. The information might be most appropriately included as part of the forest management planning documentation; in setting management objectives and targets, or in defining operating procedures (see Part Three: Using an Environmental Management System).

Where recommendations from the ESIA are incorporated into the management plan, it is important that the plan is not made too detailed, prescriptive or out of date by the inclusion of a large amount of information from the ESIA. In some countries it is a legal requirement that the forest management plan is based on an EIA.

Issues arising from the ESIA which should be addressed during planning include:

- description of expected impacts;
- description of measures to mitigate negative impacts;
- commitment to when and how they will be implemented;
- definition of responsibilities for implementing measures.

ESIA – see also:

Section 15.1: Environmental and Social Impact Assessment

Section 15.3: Writing a Management Plan

Chapter 19: Working with Stakeholders

Part 3: Using an Environmental Management System

6.2 Conservation of Biodiversity

BACKGROUND

Loss of biodiversity, through degradation and removal of forests and other natural ecosystems, is one of today's most worrying environmental problems. Tropical forests are particularly rich in plant and animal species, so their value for biodiversity conservation is high. Some tangible benefits of biological diversity include:

- a wide range of goods, services and raw materials for industry and agriculture;
- essential components of ecosystems and ecological processes on which humans depend;
- the genetic basis for breeding new varieties of food and fibre crops;
- raw genetic material for development of new medicines and drugs;
- the basis for much tourism and recreation;
- the basis for much human culture and spiritual wellbeing.

TABLE 6.2 Conservation of biodiversity

Shared FSC/ITTO requirements:	Responsibility	Scale
1 Conserve diversity at genetic, species and ecosystem levels	T/P	
2 Establish conservation zones and protected areas, including representative examples of existing ecosystems	T/P	
3 Ensure safeguards exist to protect rare, threatened and endangered species and their habitats	T	
FSC requirements:		
4 Genetically modified organisms must not be used	P	
5 Biological High Conservation Values must be maintained or enhanced	S	
6 Forest conversion to plantation or non-forest land must not occur	P	

REQUIREMENTS

1. Biological diversity is all the variation among living organisms and includes the variety within species, between species and between ecosystems. Forest management must *conserve biodiversity at all relevant genetic, species and ecosystem levels*. Conservation of biodiversity must be built into the forest management system and not regarded as an optional extra (see Section 15.6). Biodiversity is usually addressed at the:
   - ecosystem level;
   - species level;
   - genetic level.

   The forest manager must consider the effects which the silvicultural system may have on biodiversity. Many natural forest silvicultural systems selectively remove certain species from the forest, but viable populations of all species must be retained. Management guidelines should be developed and implemented to ensure the conservation of biodiversity through:
   - protection of representative forest areas;
   - protection of rare and endangered species;
   - protection of features of special biological interest, such as seed trees and keystone species (see Box 6.1);
   - identification and maintenance of biological High Conservation Values (see below).

2. The *establishment of protected areas* is an essential component of a national conservation strategy. Ideally each country should have a national network of protected areas, designated and managed by the government.

   However, the forest manager should also establish conservation zones and protected areas within the FMU as part of the conservation strategy (see Section 15.6). This is particularly important where there is little land protected nationally. Protected areas should be consistent with the national protected areas system, and should also be appropriate to the scale of the forestry operation. Representative areas of commercial forest types must also be protected. It is not sufficient to set aside inaccessible or unproductive areas such as rocky outcrops, steep slopes or

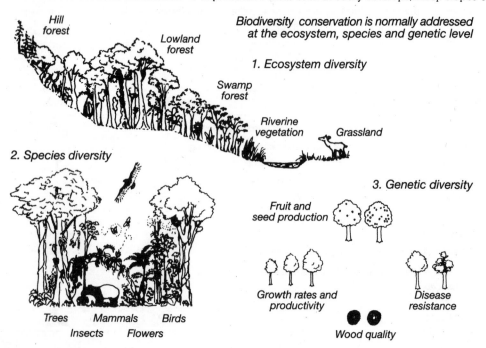

Hill forest

Lowland forest

Biodiversity conservation is normally addressed at the ecosystem, species and genetic level

1. Ecosystem diversity

Swamp forest

Riverine vegetation   Grassland

2. Species diversity

3. Genetic diversity

Fruit and seed production

Growth rates and productivity

Disease resistance

Wood quality

Trees   Mammals   Birds

Insects   Flowers

swamps. Protected areas should not be subjected to harvesting, silvicultural treatments or infrastructure development.

The size or proportion of the FMU which should be protected will depend on the local situation. For example, how much similar forest is protected outside the operation? What is the national legislation and local situation? Are there unique forest types or High Conservation Value Forests within the forest management unit? It is therefore difficult to give generalized guidelines. However, some examples of recommended or required sizes for protected areas are:

- ITTO Guidelines for the Conservation of Biological Diversity: A system of small (approximately 100 ha each) undisturbed and protected forest reserves is recommended, to act as refuges and sources for recolonization of managed forest.
- The Bolivian FSC Standard for forest management certification in the lowlands of Bolivia (September 2000) requires that at least 10 per cent of the FMU should be set aside as conservation areas (see Box 6.2).

When setting aside conservation areas, consideration should be given to fragmentation of the forest. This is when small areas of protected forest are separated by a large area of harvested forest. Animals which use the protected areas as refuges may become isolated in them if dispersal and migration links are cut. To avoid this problem, the forest manager should attempt to link together protected areas with 'wildlife corridors' of undisturbed forest. Streamside buffer zones may also provide links between areas (see Box 6.4).

Liaison with knowledgeable NGOs and academic institutions can help in the selection and management of protected areas.

Scale considerations!

Forest operations need to be over a certain size before it is considered useful to set aside conservation areas. In the case of small organizations, it is often not practical for a forest owner to set aside a significant proportion of the area for conservation. When this is the case, the forest manager should give more consideration to protection of environmentally fragile areas, and conservation of individual or small groups of old growth, habitat trees and streamside buffer zones.

Box 6.1 | **Keystone species**

Keystone species are those species upon which other animals and plants depend. For example, some animals rely heavily on one plant for a food source at certain times of year, while some plants rely on a specific animal to pollinate their flowers or disperse their seeds.

Important plant keystone species are those which produce large fruit crops at times of year when other plants do not. Fruit-eating animals may rely on these keystone species to survive a time of scarcity. Figs, for example, are thought to act as keystone species in Latin America, especially in seasonal climates – although this may not be the case in Africa. In 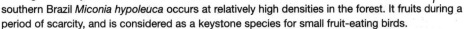 southern Brazil *Miconia hypoleuca* occurs at relatively high densities in the forest. It fruits during a period of scarcity, and is considered as a keystone species for small fruit-eating birds.

Removing keystone species, either by harvesting trees for timber, or removing them during a silvicultural treatment because they compete with a timber species, could have a devastating effect on the animal or plant populations which rely on them.

Box 6.2   **Protected areas within the forest: the Bolivian FSC example**

The FSC Standard for certification of forest management of timber yielding products in the lowlands, approved in 2000, requires that at least 10 per cent of the FMU should be set aside as conservation zones, where use is restricted.

These areas should be in different habitats, with the aim of protecting areas that are critical as refuges, feeding sites, reproduction and nesting sites of rare, threatened and endangered species. Where obvious sites for protection do not exist in the FMU, a minimum of 10 per cent of each forest type should be protected. The distribution of the sites should take into consideration the movement of wildlife through the forest, providing biological corridors and not creating 'islands' within the forest.

However, where many small landowners join together to form a cooperative or association, it may become possible, and necessary, to designate conservation zones, and compensate those owners whose land falls within such zones from the income made by the rest of the association.

3 *Rare and endangered species and their habitats*, including their nesting and feeding sites, must be protected. Basic steps towards protecting rare and endangered species include:
- Consult with local expertise, such as university experts, to determine which species are rare and endangered in the region.
- Conduct surveys, preferably linked to forest resource inventories, to establish whether rare species are present in the forest management unit.
- Develop and implement a plan to protect any rare species which have been identified.
- Monitor the performance of the species and adjust the plan if necessary.

Scale considerations! ♀

Small-scale operations are unlikely to have resources available to research and develop a full plan for protection of rare species. If rare species are known to be present in the FMU, provision for their protection during all operations should be documented in the forest management plan or operational procedures. Large operations should actively seek to establish which rare species are present and should be protected.

4 FSC-based standards do not endorse the use of *genetically modified organisms*. This does not mean that plant breeding, or the use of clones (including those produced by tissue culture techniques) is prohibited. Plant breeding and clonal populations, when used responsibly, can be an important, positive part of forest management.

5 FSC-based standards require that forest management must *maintain or enhance any High Conservation Values* (HCVs) identified in the forest. The area of forest that is needed to maintain these values is the High Conservation Value Forest (HCVF). Six HCVs have been defined (see Box 15.18, Section 15.6); four of these relate to biological values, while HCVs 5 and 6 encompass social values of the forest and are discussed in Section 7.3. HCVs can occur in any type of forest: temperate or tropical, plantation or natural.

The biological HCVs encompass exceptional or critical ecological attributes and ecosystem services. These include:
- forests that contain globally, regionally or nationally significant concentrations of biodiversity;
- globally, regionally or nationally significant large, intact landscape level forests;
- areas containing rare, threatened or endangered ecosystems;
- forests providing basic services of nature in critical situations (eg watershed protection).

Although many such areas would be protected under the normal requirements of international standards, the HCVF concept requires extra safeguards when the conservation values are of outstanding significance or critical importance.

As part of the conservation strategy (see Section 15.6), the forest manager must make an assessment of the conservation values present in the FMU and decide whether any of these should be considered as High Conservation Values. It may be possible to identify HCVs using existing information, for example, if part of the FMU has been classified as a high priority for watershed protection by government surveys, or if it borders with a Nationally Protected Area and contains similar habitat types.

Further field surveys may be needed to confirm whether or not the forest contains an HCV. Stakeholder consultation is also essential (and is required under FSC-based standards), in order to determine whether the conservation values identified are sufficiently critical or outstanding to be considered High Conservation Values.

Where HCVF is identified, specific plans must be drawn up to ensure that the HCVs are maintained or enhanced. These plans should be part of the wider conservation strategy, and need to be incorporated into overall management planning. The impact of management on the HCVs must be monitored and fed back into management at least annually.

Scale considerations! ♀

High Conservation Values may occur in any size of forest. Managers of small forests should be aware of the possibility that their forest is an HCVF and take basic steps to identify and manage any HCVs present. The process of identifying HCVF and drawing up management and monitoring plans can require specialist inputs which are beyond the capabilities of small forest owners. Small forest owners should concentrate on protecting and enhancing sites in the forest that are of clear conservation importance.

Implementing HCVF requirements in small forests is addressed specifically in Appendix 2.2. Measures the owner or manager can take to identify, manage and monitor the six HCVs are discussed and include:

- Check whether the FSC-based national standard (if available) defines HCVs for forests nationally or locally.
- Check with local university departments if they are aware of any outstanding biological values in the forest in question.
- Use personal observations and knowledge of the forest to identify any unusual sites of importance for wildlife, such as nest sites of rare species, salt licks, individual trees, hollow trees, ponds or watercourses; focus management on these sites.
- Discuss with neighbours, local community leaders and conservation groups whether they are aware of any conservation values of importance in local forests.
- If potential HCVs are identified, use common-sense measures of protection (see Box 15.23, Section 15.6) and reduced impact forest operations to try and maintain them; include the specific measures for the management and maintenance of the HCVs in the management plan and the public summary of the plan.
- Check, at least annually, that the High Conservation Value is not being adversely affected by management.

6 Under FSC-based standards, natural forests must not be converted to plantations or non-forest land. Exceptions to this are permitted only in very specific circumstances, where conversion:
  - entails a very limited portion of the FMU; and
  - does not occur on High Conservation Value Forest areas; and
  - will enable clear, substantial, secure, additional, long-term conservation benefits across the FMU.

This means that, for example, the removal of invasive trees from natural grassland areas or streamside buffer zones, comprising a small proportion of the FMU might be acceptable. However, large-scale conversion of forest to plantation or non-forest land would not be acceptable.

Conservation of biodiversity – see also:

Section 15.6: Developing a conservation strategy and the management of High Conservation Value Forests

Appendix 2.2: Implementing HCVF in small forests

6.3 Ecological Sustainability

BACKGROUND

Maintenance of ecological sustainability requires that essential resources and support processes of the ecosystem are not irreversibly disrupted. Particularly important processes and resources include:

- regeneration, succession and nutrient cycling;
- soil and water resources.

Harvesting and road construction are the forest management activities which generally have the greatest potential to impact on ecological sustainability. Particular attention must be given to controlling the impacts of these operations.

REQUIREMENTS

1 The processes of *regeneration, succession and natural nutrient cycles* are essential to SFM. Management affects these processes to a greater or lesser degree, so ways must be developed to maintain or enhance them:
- Where natural regeneration is used, consideration should be given to retaining seed trees during harvesting (see Figure 16.1). Criteria for identifying suitable trees and marking seed trees prior to harvesting should be developed where appropriate. High-grading (creaming) or taking only the best trees, and leaving defective stems, will lead to genetic degradation.

TABLE 6.3 Ecological sustainability

Shared FSC/ITTO requirements:	Responsibility	Scale
1 Maintain processes of forest regeneration, succession and natural cycles	T	🌲
2 Develop and implement guidelines for the identification and protection of sensitive soil and water resources	T	🌳
3 Prepare and implement written guidelines for road construction and use	T	🌳
4 Develop and implement reduced impact harvesting and extraction guidelines	T	🌳

| Box 6.3 | **Gaps and pioneers in natural forests** |

In a natural forest, when a tree dies, it creates a gap in the forest canopy. Increased sunlight reaches the forest floor, resulting in increased light, air temperature and soil temperature. Humidity and soil water content decrease. More nutrients are released and may be leached out of the system. The intensity of these changes depends on the size of the gap, but larger gaps experience greater changes than small ones.

These changes in conditions encourage those plants which can germinate, grow and thrive in the gaps. Some types of plants are better adapted to growing in open situations and will rapidly colonize large gaps. These species are called pioneers. Other plants can germinate and grow in shade and in these conditions will out-compete pioneer species. These 'non-pioneer' species are better suited to regeneration in small gaps. Most plants fall somewhere between these two extremes. Often, pioneer species have small, wind-dispersed seeds which blow easily into the centre of a large gap; non-pioneer species tend to have larger, heavy seeds which drop straight down and are not widely dispersed without animals to carry them.

Species which grow well in plantations are often pioneer species: pine and teak, for example. Many tropical timber species are long-lived pioneer species or light-demanding non-pioneers which germinate in shade but require considerable light to grow to the canopy. Mahogany (*Swietania macrophylla*) is adapted to large-scale disturbance, and commonly has a patchy distribution, having grown up in large gaps.

The implications for forest management are profound. Natural forest silvicultural systems are largely based on the regeneration habits of the desired species and their reaction to light. The size of the canopy gaps produced by harvesting operations in natural forest will affect the species which regenerate and therefore future crop composition.

- In natural forests, management guidelines should be drawn up for the main wood and non-wood products, covering the assessment of natural regeneration and the appropriate measures to supplement natural regeneration when necessary. Guidelines should set out the critical limits at which action must be taken to supplement natural regeneration.
- The scale and pattern of felling in natural forests should be appropriate to the natural dynamics of the forest type. In many tropical forests, the natural pattern of disturbance is caused by single trees dying and falling (see Box 6.3). Other forests are naturally adapted to more cataclysmic damage. For instance, the Mayan Forests of southern Mexico, northern Guatemala and Belize are adapted to the effects of hurricanes and fire. Where ecological requirements of the forest are not well known, it is safest to minimize canopy opening during harvesting. Large gaps, like those created by clear-felling, should always be avoided.
- Keystone tree species should be marked and protected during felling (see Box 6.1). Where commercial timber species are suspected to be keystone species, care should be taken to maintain enough trees in an area to support populations of dependent animals. Suspected keystone animal and bird species should be protected from hunting, and their habitat preserved.
- Some habitat trees may also need to be retained at harvesting. Habitat trees provide nesting space for birds and living space for insects and other animals – these may be poor quality, defective, old or very large trees. These are especially important in intensively managed forests where non-crop trees are often removed.
2 *Sensitive soil and water resources* must be protected in order to prevent damage to:
  - productivity and quality of the forest and aquatic ecosystems; and
  - downstream water quality and quantity, particularly through flooding and sedimentation.

> **Box 6.4** **Streamside buffer zones**
>
> Streamside buffer strips are strips of forest or vegetation which are left undisturbed along the banks of a watercourse or lake. Harvesting and extraction operations, road construction and sometimes establishment, are not permitted within streamside buffer strips. The objective is to protect the watercourse from sedimentation and damage to the stream banks, to prevent changes to water flows, and to provide habitat for wildlife disturbed by forestry activities.
>
>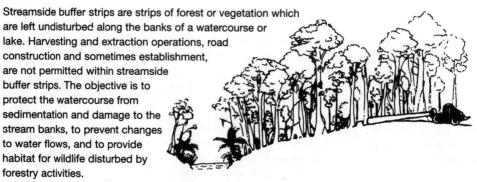
>
> Vegetation helps hold the soil and stream banks together, reducing erosion. The buffer zone also works as a sediment trap, filtering out soil particles carried in water running off roads and bare ground. The width of a buffer strip should vary according to local vegetation, soil types, topography, climate and the width of the watercourse. Where available, local guidelines should be followed. A rule of thumb is to leave a buffer of at least twice the width of the watercourse on either side; a minimum of 30m each side of a watercourse is often recommended. On steep, non-permeable soils, wider buffers are necessary. On some highly permeable soils with slopes less than 30 per cent, a 20m strip may be satisfactory.[5]
>
> In some plantation situations, such as in South Africa, streamside buffer zones are maintained as grassland to prevent excessive water use by exotic tree species.
>
> Where extraction routes have to cross a watercourse, the crossing should be located on stable materials such as gravel, or at natural fords, such as sheet rock outcrops. Crossing points should be carefully planned and marked on maps.

Guidelines or procedures should be drawn up to identify and protect sensitive soil and water resources during all operations. These should define slope limits for particular operations; soil types which need particular management prescriptions; and specifications for streamside buffer strips (see Box 6.4).

3 Road construction, maintenance and use has particular potential for damage to soil and water resources. *Guidelines to control erosion and protect water resources* during all roading operations must be developed and implemented (see Section 16.1). These should include specifications for:
   - road layout, such as maximum road density, location;
   - road design, such as road width, slopes, materials, drainage ditches and culverts;
   - road and bridge construction methods, such as equipment, labour and techniques;
   - road use and maintenance, such as rules of road use, closure after rain and frequency of maintenance;
   - road closure, such as removal of bridges or culverts and road blocks;
   - size, demarcation and protection of streamside buffer zones.

The preparation of maps is an essential part of planning roading, harvesting and extraction operations.

4 Harvesting and extraction operations have a high potential for damage to the residual stand and soil and water resources. *Reduced impact harvesting and extraction* guidelines must be developed and implemented. Practical training for supervisors and field operators is essential in order to

---

5 P F Clinnick, 'Buffer strip management in forest operations: A review', *Australian Forestry*, 48:1, pp34–45

implement such systems. Proper incentives, as well as close supervision during operations, are necessary to ensure high quality work. Reduced impact harvesting guidelines should cover:
- harvest planning;
- tree marking and felling;
- extraction; .
- post-harvest operations.

Scale considerations! 🔎

Small forest organizations are unlikely to be able to draw up their own guidelines for protecting soil and water resources, road construction and harvesting and extraction operations. Where possible, small forest organizations should make use of existing guidelines produced by government agencies, and industry or research institutions. Where these are not available, the forest owner or manager should try to put together simple guidelines for employees or contractors, and should place an emphasis on the supervision of work and on-site assessments of possible impacts (see Section 15.1) to ensure damage is minimized.

Ecological sustainability – see also:

Section 16.1: Forest Roading

Section 16.2: Harvesting and Extraction

Section 16.3: Silviculture

6.4 Use of Chemicals

BACKGROUND

Chemicals include fertilizers, insecticides, herbicides, fungicides and hormones which are used in forest management. Chemicals can have many potentially damaging effects on the environment and need careful control. Chemicals are often expensive, but good chemical practice and carefully targeted use can also reduce costs (see also Section 8.4).

REQUIREMENTS

1 *Use of chemicals should be minimized.* Current use should be reviewed to find ways to reduce the need for certain chemicals (see Box 6.5). Application of chemicals should be carefully timed to have maximum effect on the target. Fertilizers should be matched to soil deficiencies and the growing needs of plants.

   Integrated pest management (IPM) should be encouraged. This means using preventative measures, combined with curative procedures, to control pests (see Box 16.11). The aim should be to prevent pest outbreaks by good forest practice and forest hygiene measures. If preventative measures fail, silvicultural, biological and chemical control may be considered to control outbreaks.

   Extensive use of pesticides is rare in natural forest management. Natural diversity encourages natural biological control processes. However, for many forest organizations the main use of chemicals is in the nursery. The number and the quantity of chemicals being used can often be significantly reduced by a combination of good planning and nursery management.

TABLE 6.4 Use of chemicals

Shared FSC/ITTO requirements:	Responsibility	Scale
1 Minimize chemical use and adopt integrated pest management	T/S	
2 Implement procedures for handling, storage and disposal of chemicals	T	
3 Provide adequate training and equipment for chemical use	T	
4 Special restrictions must apply to chemical use in sensitive areas	T	
FSC requirements:		
5 Recognized dangerous and banned chemicals must not be used	T	
6 Use of chemical pesticides must be clearly justified and recorded	T	
7 Document, monitor, strictly control and minimize use of biological control agents	T/S	

Scale considerations!

Integrated pest management requires in-depth knowledge of the ecology of the pest species and their hosts. It is often specific to each situation, so that effective management techniques in one location may not work in another. Large-scale forestry operations should consider research into IPM: the potential cost savings and reduction in chemical use may be significant. Small forestry organizations, particularly in plantation situations, should consider preventative measures such as using appropriate planting material for the site and mixed plantings.

2 Chemical management requires *documented procedures* to ensure that all personnel in contact with chemicals are aware of the correct methods for their use. This type of information may be incorporated into procedures and work instructions, or may be included in the management plan. Procedures should cover all aspects of handling, storage and disposal of chemicals and their containers.

However carefully chemicals are stored and used, there is always some risk of accidental spillage. This risk should be considered, for example during the ESIA, and a contingency plan made which details how any spillage would be contained. It should also state who is responsible for contingency plan action.

3 In addition to good planning and management on paper, chemicals must be properly used. *Appropriate and well-maintained equipment* must be available and used, and personnel must have adequate training.

> **Box 6.5** **An example of reducing chemical use**
>
> Experiments in eastern Canada have shown that silviculture and forest management can be used to reduce incidence of Spruce budworm (*Choristoneura fumiferana*) outbreak in mixed balsam fir (*Abies balsamea*)/hardwood stands. The amount of defoliation in such stands is strongly negatively related to the hardwood percentage of the stand, especially if hardwoods make up over 40 per cent of the stand. Harvesting and silviculture could be designed to minimize budworm outbreaks.[6]

4 Chemicals must not be used in *environmentally sensitive areas* such as:
   - wetlands and watersheds;
   - streamside buffer zones;
   - conservation areas;
   - recreation areas;
   - areas near human settlements;
   - sites of special ecological significance;
   - habitats of rare or endangered species.
5 *Dangerous chemicals must not be used.* FSC-based standards prohibit use of:
   - chemical pesticides included in World Health Organization Classes 1a and 1b, including aldicarb, parathion, oxydemeton-methyl, sodium cyanide and warfarin;
   - chlorinated hydrocarbons, including aldrin, DDT, dieldrin and lindane;
   - other persistent, toxic or accumulative pesticides.
   Some temporary exceptions apply to these prohibitions. See Appendix 3.1 for a full list of prohibited chemicals.
6 Under FSC-based standards, forest managers are required to *demonstrate that they are minimizing their use of chemical pesticides*. A decision-making procedure is required for managing the use of chemical pesticides, recording and justifying why a chemical pesticide was used in preference to a non-chemical alternative. Records must be kept of all chemical pesticides used.
7 Under FSC-based standards, *biological control agents* (see Box 6.6) *may be used in preference to chemical pesticides, but must be closely controlled*. If their use is considered, the forestry organization must:
   - register and liaise with the appropriate authorities and ensure compliance with national and international laws and protocols;
   - test on a small scale before introducing generally;
   - monitor the effects on the forest fauna and flora over time.

Use of chemicals – see also:

Section 16.4: Chemicals and Pest Management

Section 15.1: Environmental and Social Impact Assessment

Section 16.5: Training

---

6 D A MacLean, 'Forest management strategies to reduce spruce budworm damage in the Fundy Model Forest', *Forestry Chronicle*, 72:4, 339–405, 1996

| Box 6.6 | **Biological control agents** |

Biological control uses natural enemies in pest control, for instance using disease-causing pathogens to control populations of pests. One group of viruses, known as nuclear polyhedrosis viruses, is particularly used, as it only affects insects and there is no risk of harming other animals or people. Biological control agents normally have to be sprayed on to a crop in the same way as insecticides, but once they begin to affect the pest population, they continue to spread, carried by the insects themselves.

The concern with biological control is the possibility that a biological control agent may itself get out of control. It is difficult to test the effects on all living organisms it may come in contact with, and once released a virus may be difficult to control. Testing before use and monitoring the effects after release are essential.

6.5 Waste Management

BACKGROUND

Forestry operations often result in considerable quantities of waste. FSC-based standards specifically cover waste management. This section covers two areas:

- managing the disposal of waste products, such as garbage, used oil, old tyres and broken vehicles;
- minimizing the waste of harvested timber, such as broken logs, large offcuts, abandoned logs, and other forest products.

REQUIREMENTS

1 *All waste should be disposed of properly.* This includes items such as litter, used oil, chemical containers, old tyres and broken-down vehicles. Where registered or recognized waste disposal sites exist within the region, these should be used. Clear procedures for the disposal of all types of waste should be understood by all employees:
   - Litter should be collected in each operational area and removed to be disposed of as appropriate.
   - Large items, such as tyres, broken vehicle parts and oil drums, must be collected from the forest and removed before operations are considered complete.
   - Oil, hydraulic fluid, fuel and other oil-based waste should not be allowed to leak on to the ground. They should be collected, removed from the forest and disposed of appropriately.
   - Used oil, chemicals and their containers should not be dumped in the forest, and should not be stored near water bodies.
2 Harvesting and processing operations can be wasteful. *Efficient harvesting minimizes waste* and reduces costs. Ways to reduce harvest waste include:
   - training fellers in directional felling techniques to reduce splitting and damage to timber and surrounding trees;
   - guidelines to fellers and cross-cutters for the optimal length to cut logs to minimize waste in the forest and at further processing stages;
   - efficient extraction and loading to ensure rapid removal of logs and reduce fungus or insect damage.
   The use of by-products and multiple purpose harvests should be encouraged to optimize use of the products harvested. For example:

TABLE 6.5 Waste management

FSC requirements:	Responsibility	Scale
1 Dispose of all waste properly, off-site where appropriate	T	
2 Minimize waste from harvesting	T	

- In some forests larger stems are used for sawn timber, medium-sized stems for pulp and small diameter stems for charcoal.
- In natural forest harvested for plywood, only logs of a certain length can be used. Cut-off log ends and peeler cores may be used to make furniture.
- If high quality logs are sent for export, logs with small defects can be used for local markets or processing – rather than being abandoned.
- Epiphytes, such as bromeliads, growing on felled trees may be harvested and sold.

Waste management – see also:

Section 15.1: Environmental and Social Impact Assessment

Section 16.4: Chemicals and Pest Management

Section 16.2: Harvesting and Extraction

7 The Wellbeing of People

At its best, forestry can help communities raise their standard of living and contribute to positive development. At its worst, it can destroy species and landscapes that people value, throw people off land which they consider to be their own, and degrade the relationships which keep communities and societies together.

Some forestry operations have come to suffer financially from the social problems they cause. Most of the worst conflicts in forestry today are between companies and local communities. Equally, other companies are increasing the long-term security of their investments by entering into productive relationships with local people.

Some social issues need to be addressed by the forest management team directly, while others may require specialist knowledge. Some responsibilities rest with the government.

Many social issues are contentious and easy solutions do not exist. Often the forest manager can only attempt to find the best balance possible between conflicting demands on the forest resource. FSC-based standards particularly emphasize the social responsibilities of forestry organizations.

This chapter considers the need for:

- consultation and participation processes;
- social impact assessment;
- recognition of rights and culture;
- relations with employees;
- contribution to development.

More information about working with people can be found in Part Five.

7.1 Consultation and Participation Processes

BACKGROUND

Stakeholders are all the people who are interested in, or affected by, forest management and operations. Stakeholders may include:

- government agencies;
- communities in and around forested areas;
- indigenous people;
- employees and contractors;
- investors and insurers;
- customers and consumers;
- environmental interest groups;
- general public.

Different stakeholders may have different objectives and interests, knowledge, skills, rights or influence in affected forests. Stakeholders' potential contributions to good forest management vary: some may seem only to cause trouble, but many are crucial to the long-term success and sustainability of the forest organization. As a result, consultation and participation processes are

TABLE 7.1 Consultation and participation processes

Shared FSC/ITTO requirements	Responsibility	Scale
1 Maintain consultations and encourage participation by people and groups affected by forest operations	T	🌲

FSC requirements

2 Employ appropriate mechanisms for resolving grievances and providing compensation	P/T	🌲

now recognized as central aspects of SFM. However, it is important that consultation and participation is done sensitively and realistically. It should not raise expectations which are beyond the ability of the forestry organization to meet. Where concerns are raised which are properly the responsibility of government or other bodies, good communications are key to resolution. Specialist help may also be necessary.

REQUIREMENTS

1 *Consultation* is the two-way flow of information between forest managers and other stakeholders. Consultation is more than just listening to stakeholders: it should also involve the participation of stakeholders in the joint planning of relevant management decisions. If consultation is carried out regularly, stakeholders should understand each others' plans and interests, allowing problems to be resolved fast.

One problem forest managers may face is a history of misunderstanding, resentment and bad feeling between stakeholders and their organization. If this is the case, training to help forest management staff understand the rationale for developing a stronger consultation and participation process may be necessary.

Firstly, all potential stakeholders must be identified. Then a process is needed for distinguishing which stakeholders have a legitimate interest or key contribution to make – these are the stakeholders which forest managers need to work with. Ideally this should be done before operations begin.

On-going consultation is often achieved by forming a working group, liaison committee or local board. This should consist of forest managers, key staff, local people and other key stakeholders. The functions of this group are to:
- bring up worries or complaints before people become frustrated;
- develop a partnership through regular face-to-face exchange;
- review monitoring reports.

The consultation process should be documented and records of attendance, meetings and decisions made publicly available. A *summary of the management plan and monitoring results should also be accessible to all stakeholders.*

2 Appropriate mechanisms are also needed for the *resolution of grievances* and to provide fair compensation for damage or loss affecting tenure rights, property, resources or livelihoods. As with regular, on-going consultation, a local board may be the best way to address this. The board must agree the rules by which it will operate and must keep records of who attends and what happens in its meetings. The board must maintain a reputation for fairness and impartiality in the resolution of grievances. In some countries there are recognized procedures for dealing with grievances that take account of local norms, and these should be used if they are widely accepted.

Consultation and participation processes – see also:

Section 10.1: Developing Commitment

Section 15.2: Communication and Collaboration with Stakeholders

Chapter 19: Working with Stakeholders

7.2 Social impact assessment

BACKGROUND

Social impact assessment is the social equivalent of environmental impact assessment (EIA) – and for this reason it is often carried out at the same time. A social impact assessment involves collecting information about stakeholders and their interests, and then considering the actual, or possible, impacts of forest operations on these interests in order to propose mitigation measures. A social impact assessment requires extensive consultation with stakeholders. It can therefore be a good means for developing consultation and participation among stakeholders.

REQUIREMENTS

1 *Social impact assessment* should generally be combined with environmental impact assessment in an ESIA, because stakeholders will raise environmental concerns too. Ideally, an ESIA should be carried out before operations start.

Social impact assessment should include the following processes:
- identification of stakeholders and their interests;
- identification and evaluation of potentially significant social impacts;
- consideration of options for avoiding or mitigating negative impacts and optimizing positive impacts;
- prioritization of activities required to address impacts;
- incorporation of activities into the management planning process.

Scale considerations!

To produce a full social impact assessment for a medium to large-scale operation requires specialist inputs. Badly executed social impact assessment runs the risk of incorrectly identifying impacts and potential problems. It could also incorrectly portray the forest organization's aims to stakeholders, which may result in wrong expectations, misunderstandings and unnecessary conflicts.

Small-scale operations with low impacts that cannot bring in specialist advice should try to:

- discuss planned management with neighbours and other users of land, especially regarding questions of access to land and forest resources;
- discuss with local or regional authorities, particularly regarding use and maintenance of roads to be used by the forest organization;
- hold an open meeting with interested parties to permit open discussion of concerns and ideas;
- maintain a record of discussions, including participants, dates, issues discussed and conclusions;
- incorporate results of discussions into management planning documentation, noting justifications.

TABLE 7.2 **Social impact assessment**

Shared FSC/ITTO requirements:	Responsibility	Scale
1 Carry out social impact assessment and incorporate results into management planning	S	
2 Take measures to avoid possible negative social impacts	T	

2 Once possible negative social impacts have been identified, *steps must be taken to avoid them.* The planning and assessment of alternative activities should be considered as part of the social impact assessment; where alternatives are not possible, mitigation measures to reduce impacts must then be implemented. Social targets and responsibility for social issues must be clearly defined in the management planning documentation.

Social impact assessment – see also

Section 15.1: Environmental and Social Impact Assessment

Section 15.2: Communication and Collaboration with Stakeholders

Section 19.5: Agreeing Social Responsibilities

Chapter 19: Working with Stakeholders

7.3 Recognition of Rights and Culture

BACKGROUND

The issues of rights and culture are at the core of the wellbeing of stakeholders. The obligation of the forest organization to recognize and respect legal tenure and use rights was covered earlier (see Section 4.2). This section discusses the necessity of respecting traditional and indigenous use and access rights. These may not be detailed or even recognized in national law, but may be critical to local livelihoods and culture. FSC-based standards require that forest management must maintain or enhance any High Conservation Values (HCVs) identified in the forest (see Section 15.6). Two types of social HCVs are recognized. These include areas of forest that are fundamental to meeting the basic needs of local communities and those areas critical to local communities' cultural identity.

REQUIREMENTS

1 Local communities and indigenous people with *legal or customary tenure or use rights* should be able to maintain these rights with respect to any forest management operations on their land. All legal or customary tenure or use rights to the forest must be clearly documented and mapped by forest managers. Forest management activities must not threaten or diminish the resource rights, use rights or tenure rights of local and indigenous communities. Such communities may use external

TABLE 7.3 Recognition of rights and culture

Shared FSC/ITTO requirements:	Responsibility	Scale
1 Recognize and uphold legal and customary rights of local and indigenous communities to control management on their lands	P	
2 Identify and protect sites of special cultural, ecological, economic or spiritual significance to indigenous peoples	T	

FSC requirements:		
3 Compensate indigenous peoples for the application of their traditional knowledge	P	
4 Social High Conservation Values must be maintained or enhanced	T	

specialists in negotiating deals to exercise these rights over resources if required, but all cases of delegation of consent must be shown to be free and informed – based on a free flow of information to all concerned people. The process of consultation and negotiation must be documented.

Since plantations may involve a significant change in the way the land is used, special attention should be paid to protection of local rights of ownership, use or access in land acquisition for plantations.

2 Sites of special *cultural, ecological, economic or spiritual significance to indigenous peoples must be protected* from damage by forestry operations. A system is needed to identify, record and protect such sites, in cooperation with local and indigenous people. These sites should be identified on maps, and protected in a manner similar to conservation zones, as discussed earlier (see Section 6.2).

3 Compensation for the application of indigenous people's *traditional knowledge* must be agreed before operations utilizing that information commence. Compensation levels must be set with the free and informed consent of indigenous peoples and full awareness of the potential value of their knowledge.

| Box 7.1 | **Compensation for traditional knowledge** |

Tropical forests represent a huge storehouse of biodiversity: genetic and biochemical resources of particular importance to the pharmaceutical, agricultural and biotechnological industries. Indigenous people often hold clues about which animals and plants may hold potential for what uses, based on their traditional use of the forest. However, the value of this knowledge and compensation for its commercial use is rarely acknowledged. In 1985 the market value of plant-based medicines sold in developed countries was US$43 billion. It is estimated that less than 0.001 per cent of the profits were returned to the indigenous peoples from whom much of the original knowledge came.[7]

Benefit-sharing and prior informed consent of indigenous communities is now required under a range of international policy instruments.[8] An international access and benefit-sharing regime is being negotiated under the auspices of the Convention on Biological Diversity as mandated by the World Summit on Sustainable Development in 2002.

This is particularly important where a forest organization is considering commercial harvests of traditionally utilized non-timber forest products and medicinal plants with potential for use in the pharmaceutical, biotechnology and agricultural industries (see Box 7.1). In such situations, indigenous peoples must be compensated for the fair market value of their traditional knowledge.

4 FSC-based standards require that forest management must *maintain or enhance social High Conservation Values* (HCVs) identified in the forest. Six HCVs have been defined (see Box 15.18, Section 15.6); HCVs 5 and 6 relate to social values of the forest, while the other four are biological HCVs and are discussed in Section 6.2.

The social HCVs encompass:
* areas of forest that are fundamental to meeting the basic needs of local communities (such as subsistence or health);
* areas that are critical to local communities' traditional cultural identity.

While the identification and protection of sites of special significance for indigenous people is an important general requirement of international standards, the management of social HCVs should identify and protect the areas of forest that are of critical or outstanding importance to local communities.

Social HCVs must be identified in cooperation with local people to whom they may be important. This means proactively seeking the views of local communities through a consultation process (see Section 15.2), and deciding with them which values (if any) of the forest should be considered *High* Conservation Values. The management regime of any social HCVs should be agreed with the local communities to whom they are important.

The identification, management and monitoring of HCVs is dealt with in more detail in Section 15.6. Specific management measures aimed at ensuring that forest management maintains or enhances social HCVs should be detailed in the management plan and the public summary of the plan. The impact of management on the HCVs must be monitored and fed back into management at least annually. Participatory monitoring (see Chapter 21) may be particularly appropriate in the case of social HCVs.

---

7 D Posey, 'Indigenous knowledge, biodiversity and international rights', *Commonwealth Forestry Review* 76(1), March, 1997
8 S Laird, 'Ethical and legal obligations of commercialising traditional knowledge and resources'. In P Shanley et al, *Tapping the Green Market*. Earthscan, London, 2002

Recognition of rights and culture – see also:

Section 15.1: Environmental and Social Impact Assessment

Section 15.2: Communication and Collaboration with Stakeholders

Chapter 18: Why Social Issues are Important

Section 19.5: Agreeing Social Responsibilities

7.4 Relations with Employees

BACKGROUND

Forest workers are important stakeholders in forest management. Sustainable forest management is not possible without workers being capable and willing to work efficiently, avoiding damage to trees, to the environment, to equipment, to other people or themselves. All staff should be satisfied with their job and willing to continue doing it.

However, in many countries forestry work is insecure and underpaid, with insufficient safety provisions. It is also hard and dangerous work with a high risk of injury. Typically this means an inefficient and unstable workforce and high long-term costs for forestry operations – in spite of low wages. It may also mean management faces many labour problems. Sustainable forest management depends on the sustainable management of human resources as well as natural resources.

REQUIREMENTS

1 The forest organization must meet, or exceed, all applicable laws and regulations covering *health and safety of employees and their families*. Where health and safety regulations have not been issued by government, appropriate rules and procedures must be developed by the forest organization. Employers must also ensure that workers are adequately instructed in safe working techniques and are provided with safe equipment. Matching the right job to the right worker is crucial. This involves consideration of body size, age, health and nutrition, as well as education and training.

As well as risk from accidental injuries, in some countries forest workers are also at risk from illnesses such as malaria, HIV/AIDS, digestive and respiratory diseases and worm infestation. Employers should provide preventative and curative medical assistance where public services are inadequate or unavailable. Emergency evacuation procedures must be in place, in case someone has a serious accident or illness which requires immediate hospital treatment.

2 Operations which may have a direct effect on sustainability, such as planning, tree felling, road construction and skidding, must be done by staff with adequate training. Training is essential at all levels in an organization to ensure that there is capacity to carry out planning, implementation and monitoring of SFM. As well as increasing skill levels, training also provides a motivating force and helps build confidence and improve performance.

At the minimum all staff (including contractors – see Box 7.3) must receive sufficient training to ensure that they understand:

• all technical aspects of their work;

TABLE 7.4 Relations with employees

Shared FSC/ITTO requirements	Responsibility	Scale
1 Meet or exceed all applicable laws and/or regulations covering health and safety of employees and their families	T	
2 Provide adequate training for all staff	T	
FSC requirements:		
3 Provide opportunities for employment and training to communities in or near the forest operation	P	
4 Guarantee the rights of workers to organize and negotiate	P	
5 Comply with applicable conventions of the International Labour Organization	T	

- how to achieve environmental and social objectives in relation to this work and why these are important;
- how to work safely and what to do in the event of an accident.

Training needs analysis in larger forest organizations should be incorporated into routine staff management and development programmes. A training needs analysis will help identify the gap between what staff need to be able to do, and what they are currently trained to do. Records must be kept of the training each staff member requires, what training they have already received and when further training is planned.

3 Employment is the most common local benefit of a forestry operation. Job opportunities should be provided for *employment of local people at all levels* within the organization. Where suitably qualified local people are not available, the forest organization must consider means of providing appropriate training for local people to benefit from employment opportunities.

Box 7.2 Addressing common problems with health and safety issues

Where there is a lack of awareness of health and safety and environmental issues among field staff consider trying to:

1 Convene a safety committee involving representatives from all levels of the forest organization so that staff feel as involved as managers.

2 Hold short (five minutes) daily or weekly talks in the field before starting work ('tool-box talks') covering a different health and safety or environmental issue at each.

3 Develop pride in different departments in their safety or environmental records by scoring or ranking departments' performance every month.

In addition to creating jobs within the forest organization, skills development and support for jobs or businesses in the community will reduce reliance of the community on the individual forest organization.

4 Forest organizations must guarantee workers' rights to trade union representation and collective bargaining in accordance with the Conventions of the International Labour Organization.

5 There are some 15 Conventions of the International Labour Organization (ILO) that have an impact on forestry operations and practices and these are mandatory in almost all countries. Details of the ILO Conventions and their implications for forestry are given in Appendix 3.3.

Forest organizations should pay decent wages in relation to the general level of wages in the country and must comply with minimum wage legislation. Furthermore, forest organizations must not employ children, defined as those under the age of completion of compulsory schooling (normally not less than 15 years). Forest organizations must not discriminate in employment on the basis of race, colour, sex, religion, political opinion or origin.[9]

Relations with employees – see also:

 Section 15.2: Communication and Collaboration with Stakeholders

 Section 16.5: Training

 Chapter 20: Conditions of Employment

Box 7.3 | **Addressing common problems with training and conditions amongst contractors**

In many countries, forest organizations are increasingly moving away from using directly employed labour to contracting out operations to contractors. This has often resulted in lower quality jobs in terms of income, job quality, working hours, health and safety and social security coverage. Training and other conditions of service are often regarded as the responsibility of the contractor.

In terms of SFM, however, the training requirements and conditions of contract staff and directly employed staff are the same. In order for operations to be adequately planned and implemented, managers and operators must receive sufficient training and supervision. They must also be adequately paid and work in decent conditions (eg have adequate personal safety equipment, appropriate tools for the job, food and housing facilities). Means of assisting contractors and ensuring that they use qualified and trained staff include:

- maintaining a list or register of approved contractors and their staff, with copies of required training certificates, etc;

- performing spot checks on contractors' employees to check that working conditions are decent and only trained staff are operating;

- providing assistance for training courses for contractors, such as by providing sites, or part funding for courses;

- working through established contractors' organizations and associations.

---

9 'FSC Certification and the ILO Conventions'. FSC Policy Paper, Forest Stewardship Council, Bonn. 2002. Available at www.fscoax.org

P Poschen, *Social Criteria and Indicators for Sustainable Forest Management: A Guide to the ILO Texts*. International Labour Organization, Geneva 2000

TABLE 7.5 **Contribution to development**

Shared FSC/ITTO requirements:	Responsibility	Scale
1 Contribute to an equitable distribution of the benefits, costs and incentives of forest management	P	🌲
FSC requirements:		
2 Strive to strengthen and diversify the local economy	P	🌲

7.5 Contribution to Development

BACKGROUND

A forest organization alone cannot be expected to achieve the goals of wider society, such as equitable rural development. Such goals are primarily those of governments and are affected by factors that are beyond the control of the forest organization.

However, forest management which incorporates social issues effectively can play an important role in local development. Forest organizations can improve human wellbeing in forested areas – and should work to avoid inequitable access to the benefits of forestry.

REQUIREMENTS

1 The forest organization should contribute to an *equitable distribution of the costs, benefits and incentives* of forest management, between the owners of the resource, the forestry organization and local communities. In part, these obligations are set out in legislation and may include payment of royalties, taxes, development funds and other levies to national and local institutions.

Benefit-sharing is not necessarily a monetary transaction. Services which can be provided by the forest organization, such as transport, medical facilities or schooling, may be important to local communities. The best means of contributing to local development should be negotiated with the communities in question through the process of consultation. Possible ways of benefit-sharing are described in Section 19.5.

2 Forest managers should *encourage the development and diversification of local forest enterprises*. This may be based on local processing of wood products from the forest operation, or on processing and marketing NTFPs. Development of non-traditional products can also prevent over-dependence by the local economy on one, or only a few, products (see also Section 5.5).

For example, Mondi Forests in South Africa encourages local small businesses to buy waste log ends which are then turned into tomato boxes. In some areas forestry companies encourage local communities to grow crops, such as groundnuts, which fix nitrogen in the soil, between newly planted rows of seedlings. The communities benefit from having access to additional land to grow food or cash crops and the forestry companies benefit from improved weed control, soil fertility and better fire protection.[10]

In Ghana the Forestry Department successfully introduced free permits for gathering *Marantaceae* leaves. These leaves, most commonly found growing in tree-fall gaps, are widely used as wrappers for roadside sales of fried plantain and cooked food as well as for wrapping cola nuts, spices, salt and meat. For many villagers, and especially women, selling these leaves can be an important source of income.[11]

10 *Benefits and Costs of Plantation Forestry: Case Studies from Mpumalanga.* CSIR for Department of Water Affairs and Forestry, South Africa, 1995
11 M Agyemang, *The Leaf Gatherers of Kwapanin, Ghana*, Forestry Participation Series No 1 IIED, London, 1996

Contribution to development – see also:

Section 15.2: Communication and Collaboration with Stakeholders

Section 19.5: Agreeing Social Responsibilities

8 Plantations

Plantations and substantially planted forests can contribute a range of benefits to society but they rarely produce all the social and environmental benefits of natural forests. Planting trees will also induce changes in the local biological and physical environment which may have negative social and environmental impacts.

Plantations should be planned and designed to complement the management of neighbouring natural forest. They should be integrated with other land uses to ensure conservation of native flora and fauna, maintenance of ecological sustainability and maximization of social benefits, while providing a productive resource.

This chapter covers the issues of specific importance to plantation management. However, it is designed to complement, not replace, other chapters of the handbook. Topics include:

- plantation planning;
- species selection;
- soil and site management;
- pest and disease management;
- conservation and restoration of natural forest cover.

8.1 Plantation planning

BACKGROUND

Plantations have a significant impact on the landscape and local biodiversity. Plantation planning needs to take into account the wider context in which the plantation is located, the effects it will have on the ecology and landscape, and the contribution to, or adverse impacts on, local biodiversity.

REQUIREMENTS

1 Plantations should be planned within the context of local land use. Environmental and social impact assessment and management planning should *consider the impacts on local communities, conservation of biodiversity and the landscape.* The layout of the plantation should be designed to protect areas of natural forest within the forest management unit (see Section 8.5) and promote protection of forests outside the forest management unit.

   In countries where visual aspects of the countryside are important for local cultural values or tourism, plantation layout should fit with the landscape.
2 In the case of planning for new plantations on land acquired through purchase, lease or other arrangement with the landowners, particular *attention must be paid to social issues.* Forest managers investing in planting schemes generally want sole rights to the land, but this is not always possible.

   In many countries, rural land was originally owned by communities but may have been appropriated in the past on behalf of the state. This land may later be given or sold to third parties, such as plantation investors. Forest dwellers and community members often argue that governments have seized their land illegally and their claims may be backed by some national and

TABLE 8.1 Plantation planning

Shared FSC/ITTO requirements:	Responsibility	Scale
1 Consider impacts of plantation on local communities, biodiversity and landscape: design and layout to promote protection of natural forests	P/T	
2 Pay special attention to social issues of land acquisition for plantations	P	
3 Design layout to include a mosaic of stands	T	
ITTO requirements:		
4 Carry out detailed site survey as the basis of planning	T	
5 Carry out integrated resource inventories to evaluate risks and opportunities of plantation	T	
6 Before final harvest, plan for following generation	T	

international laws. There may also be other use rights and histories of claims by different interests which must be dealt with. Social impact assessment is an effective tool to begin tackling these issues (see Sections 7.2 and 15.1).

3 The plantation layout should be designed to include as much diversity as possible within productive areas. The principles of ecological landscape planning (see Box 8.1) should be applied to plantation layout. This includes:
- creating variety in the species composition, size and distribution of management blocks;
- creating a mosaic of stands, combining protected old-growth or natural stands, in close proximity to logged stands;
- maintaining wildlife corridors of protected forest within the plantation, connecting old-growth forest or areas of natural vegetation; this can include the use of streamside buffer strips, roadside fringes of native species, and upland corridors crossing ridges and connecting valleys;
- planting local, native species as a fringe around the edges of a plantation, or individual blocks;
- establishment of stands with different rotation periods within the forest operation.

Scale considerations!

Small-scale organizations may not be large enough to include a mosaic of stands within the forest operation. The forest manager should aim to complement the landscape outside the forest management unit and add diversity within the stand by managing individual old-growth trees, streamside buffer zones and wildlife corridors.

4 ITTO guidelines require that *detailed site surveys* should be carried out as the basis of all planning. The site survey should provide information to:
- identify land available for planting or other uses, such as conservation zones;
- plan plantation layout and design;
- determine soil and site classification for species selection;
- carry out site classification to determine preparation methods;
- identify soil nutrient status to determine soil fertilization or amelioration requirements.

5 *Integrated resource inventories* should be conducted of plantation resources to provide information on:

Box 8.1 **Ecological landscape planning**

Natural forest ecosystems are characterized by variety: many species, irregularity of spacing and shapes, a range of ages, and so on. This increases the diversity of the forest. Ecological landscape planning aims to mimic this diversity, to provide as natural as possible an ecosystem. Three considerations are important for maintaining forest landscape ecology:

1 connecting corridors of protected forest, such as streamside buffer zones and upland corridors;

2 ensuring a variety of species composition, distributions and sizes of stands;

3 planning a variety of rotation lengths, including extended rotations of stands, and retention of individual trees or small patches of forest within the stand.[12]

- the health of the forest ecosystem and environment and any significant risk factors;
- the state and potential development of biodiversity;
- the opportunities for wildlife conservation and management;
- the quality and assortment of types and sizes of the timber resource;
- the opportunities for recreation and production of NTFPs and other services.

This information should be used to guide alternative management options, to optimize the benefits from the forest.

6 Before final harvest of a stand, the *following generation should be planned*. In particular, opportunities for increasing biodiversity within the stand should be maximized by retaining:

- live trees, especially large trees and those with features such as rot pockets, cavities, and large limbs;
- standing dead trees in varying stages of decay;
- logs and woody debris;
- undisturbed layers of forest floor;
- forest understorey plants such as moss, herbs, shrubs and small trees.[13]

Plantation planning – see also:

Section 15.1: Environmental and Social Impact Assessment

Section 15.3 Writing a Management Plan

---

12 For more information see H Hammond, *Seeing the Forest Among the Trees: The Case for Holistic Forest Use*. Polestar Book Publishers, Canada, 1991

13 Taken from JF Franklin, DR Berg, DA Thornburgh and JC Tappeiner 'Alternative silvicultural approaches to timber harvesting: Variable retention harvest systems', in *Creating a Forestry for the 21st Century: The Science of Ecosystem Management*, ed. KA Kohm and JF Franklin, Island Press, Washington, DC, 1996

8.2 Species Selection

BACKGROUND

Selection of species, provenances and clonal material should be based on their known suitability for the site and their ability to fulfil the management objectives. This should be balanced with a consideration of the advantages of retaining some diversity of planted material.

REQUIREMENTS

1 Selection of planting stock must be based on suitability to the site conditions and the management objectives. This includes *selection of species, provenance and genotype*. Choice of suitable planting stock should be based on:
   - soil and site survey;
   - management objectives and desired products;
   - market trends;
   - performance in local trials and experience with similar sites;
   - resources available for management, including time and labour;
   - social and cultural values, such as local landscape preferences and use of tree by-products, for example firewood and fodder;
   - climate;
   - ecological risks, such as storms, flooding, pests and diseases, drought and fire.

   It is essential that records are kept of the source of the material, including species, provenance, clone or hybrid information.

2 *Native species are preferable* as planting stock, in terms of their contribution to conservation of biodiversity (see Box 8.2). Native species may also have adaptations to local environmental conditions which will give them long-term advantage over exotic species. They may also require fewer management inputs. Exotic species should only be used where they out-perform native species in documented trials or local experience.

3 Species selection should *favour diversity at all scales*. This includes variety between forest management units (see Section 8.1) and variety within the stand, including diversity of species, genotypes, and ages of individual trees.

4 *Research is essential* to determine suitability of species to the site. Widespread planting should not be carried out before local trials or experience have shown that a species is ecologically well adapted to the site and is not an invasive weed outside the plantation.

   Some plantation species are aggressive colonizers of open land. As such, there is a danger of them becoming weeds, regenerating in farmland or invading natural forests. It is also important to ensure that aggressive provenances of native species, which do not come from the local area, do not become invasive weeds, as this can lead to destruction of local provenances and loss of genetic diversity.

TABLE 8.2 Species selection

Shared FSC/ITTO requirements:	Responsibility	Scale
1 Select planting stock based on appropriateness to the site conditions and management objectives	T	
2 Native species should be used in preference to exotics	T	
3 Diversity is preferred at all scales	T	
4 Carry out research to determine suitability to site	T/S	

| Box 8.2 | **Exotic and native species** |

An exotic species is one which is not naturally found in the area in question; it has been introduced from elsewhere. Species such as *Eucalyptus* spp, *Acacia* spp, or *Pinus* spp are often used in plantations outside their natural range because:

- The biology of widespread exotics such as these is well known. Seed of well-documented origin is available, nursery practices and silviculture are well defined and results are fairly predictable.

- For some sites, no native species may be suitable, but an exotic tree may tolerate the conditions. For example, *Acacia mangium* is widely planted because of its tolerance of a variety of environmental conditions such as flooding, acid soil, burning and competition from grasses.

- Exotic species newly introduced to an area may be free of their native pests and diseases, at least temporarily, allowing vigorous growth.

Some drawbacks with using exotic species include:

- They support less diversity of wildlife so have a lower conservation value than native trees.

- They sometimes suffer greater attacks from local pests because they have not developed natural defences.

- Exotics may turn into weeds in areas where they have no natural competitors, pests or diseases. Plantation species are pioneer species and may colonize open farmland. For example, in many areas of South Africa, black wattle, *Acacia mearnsii*, which was introduced from Australia, has become a common weed.

Systematic research should also be carried out on native species which have not been tested in plantation conditions. Native species which apparently are out-performed by exotics may perform better when grown in plantation conditions.

Scale considerations!

Research into species, provenance and clonal suitability to a site requires considerable effort and statistical analysis over several years. Small forestry organizations are unlikely to have expertise and time consistently available to implement such research. They should make use of relevant results from research projects and institutes in the local area. Large organizations should develop their own research capacity, coordinated with research institutions.

Species selection – see also:

Section 15.1: Environmental and Social Impact Assessment

Chapter 17: Monitoring

8.3 Soil and Site Management

BACKGROUND

Plantations are generally an intensive form of forestry. As such they may make higher demands on soil and water resources than other forms of forestry. Plantation management often involves extreme changes to the environment, particularly at establishment and felling phases. In particular:

- soil may be exposed to erosion;
- nutrients may be leached out;
- water quality, quantity and drainage patterns may be disturbed.

Proper management of soil and water resources is essential for the success of the plantation and reduces the risk of unwanted off-site effects.

REQUIREMENTS

1 *Protection of soil resources* is essential for the sustainable management of plantation forests. Soil flora, fauna and microbe activity is necessary for maintaining soil fertility and structure. These essential processes are especially vulnerable when soil is left unprotected by vegetative cover. Measures should be taken to minimize soil damage, by:
   - utilizing soil cover crops, especially leguminous crops which add nitrogen to the soil;
   - establishing and maintaining a diverse understorey;
   - avoiding use of heavy machinery on fragile and waterlogged soils.
2 *All watercourses and riparian areas must be protected* by adequately wide streamside buffer strips (see Section 6.3 and Box 6.4). These are intended to maximize absorption of overland water flow, nutrients and sediments from disturbed sites. It is particularly important that streamside buffer strips are respected during roading, site preparation, establishment and harvesting phases when soil is likely to be exposed and vulnerable to erosion and run-off.
3 *Low-impact weed control methods* should be used in preference to chemical control or intensive mechanical disturbance. This may include:

TABLE 8.3 Soil and site management

Shared FSC/ITTO requirements:	Responsibility	Scale
1 Minimize soil exposure and maintain or improve soil structure and fertility	T	
2 Establish completely protected streamside buffer zones along all watercourses	T	
3 Use low-impact weed control methods	T	
4 Restrict applications of fertilizer	T	
ITTO requirements:		
5 Restrict intensive site preparation to suitable sites	T	
FSC requirements:		
6 Maintain water quality, quantity and drainage patterns	T	

- Leaving non-tree crop vegetation on site to promote ecological stability and provide ground cover and soil fertility, such as legumes, which fix nitrogen in the soil. For example, in some areas of Mpumalanga, South Africa, forestry companies allow local communities to grow groundnuts between newly planted rows of tree seedlings.[14]
- Manual cutting or slashing of invasive weeds.
- Controlled burning.

4 *Applications of fertilizer should be minimized* and restricted to sites where it is not likely to be transported to streams, waterways or groundwater. Inappropriate use of fertilizer can lead to a variety of environmental problems from accumulation of heavy metals in the soil to eutrophication of water bodies (see Box 16.10). Priority should be given to use of organic and biological fertilization methods. Land capability assessment will assist in the determination of fertilization specifications.

5 ITTO Guidelines require that site survey and mapping should be used to classify the forest management unit according to *site risk and production potential*. Forest crops which require intensive site preparation, normally associated with short-rotation, intensive, industrial tree plantations, must be restricted to sites with favourable soils and flat or gently sloping terrain.

Unfavourable sites such as steep slopes, and fragile or deficient soils, are high risk and low productivity areas. They should be reserved for protection and conservation forestry.

6 *Water quality, quantity and drainage patterns* can all be affected by afforestation. Where there is a danger that water resources could be diminished or damaged by forestry activities, technical advice should be sought, as the causes and effects are very variable. For example, in the UK, plantations of sitka spruce (*Picea sitchensis*) have been found to increase acidity of water running off the plantations.[15]

FSC-based standards require that:

- Water quality must be protected by the maintenance of streamside buffer zones along all watercourses. Water running off a plantation may carry fertilizer, pesticides, oil or fuel, and sediments from soil erosion.
- Water yields must not be affected by establishment of plantations. For example, in South Africa, plantations of exotic trees in grasslands can result in reduced mean annual run-off, causing low base flows in winter. In this case, wide, unplanted buffer zones around watercourses help protect water yields.
- Drainage patterns must not be significantly altered by plantation activities. This is especially important where natural wetlands of biodiversity importance, or surrounding agricultural lands, may be affected by changed drainage patterns.

---

14 *Benefits and Costs of Plantation Forestry, Case Studies from Mpumalanga*, CSIR for the Department of Water Affairs and Forestry, South Africa, 1995

15 *Nature Conservation and Afforestation in Britain*, Nature Conservancy Council, 1986

Soil and site management – see also:

Section 16.2: Harvesting and Extraction

Section 16.4: Chemicals and Pest Management

8.4 Pest and Disease Management

BACKGROUND

Extensive plantations of genetically similar material are vulnerable to attack by pests and diseases. In natural forests, predators and parasites tend to keep pests in check; natural diversity slows the spread of diseases. Incorporation of diversity in species, provenances, genotypes and ages, into plantations usually increases resistance to pest and disease attacks.

The following requirements have been largely covered elsewhere and are cross-referenced here.

REQUIREMENTS

1 *Species, provenance and genotype must be appropriate to the site* and management practices. Healthy, well-adapted, growing stock has a natural resistance to pests and diseases. Careful species selection, based on systematic research to determine suitability to local conditions, is essential (see Section 8.2).
2 *Use of chemicals should be minimized* and carefully targeted.
3 *Integrated pest management* systems and forest hygiene practices should be developed to minimize use of chemicals (see section 6.4).

Pest and disease management – see also:

Section 16.4: Chemicals and Pest Management

TABLE 8.4 **Pest and disease management**

Shared FSC/ITTO requirements:	Responsibility	Scale
1 Carefully match species to site and cultural practices to prevent pest and disease outbreaks	T	
2 Avoid indiscriminate, blanket applications of chemicals; aim to reduce chemical application in general	T	
3 Use integrated pest management	S	

8.5 Conservation and Restoration of Natural Forest Cover

BACKGROUND

Many plantations are composed of only one or a few tree species, which provide poor habitat for wildlife and contribute little to conservation of biodiversity, especially if they are exotics. The FSC-based standards emphasize the importance of actively managing plantations to maintain or restore some natural forest cover and thereby increase the biodiversity conservation value.

REQUIREMENTS

1 FSC-based standards require that *natural forests must not be replaced by tree plantations or other land uses*. Conversion may only be considered acceptable where it occurs on a very limited portion of the FMU, does not occur in HCVF and will enable clear, additional, secure and long-term conservation benefits. Plantations should be planned within the landscape to complement the use of, and reduce pressure on, natural forest resources. Trees planted within natural forests must not replace or significantly alter the natural ecosystem.

2. A proportion of the overall forest management area must be actively managed for the *restoration of natural forest cover*. The proportion should be determined by regional standards and should be appropriate to the scale of the forest operation.

   If a proportion of the area is already covered in natural forest, this must be maintained and managed for native biodiversity conservation. Areas of remaining natural forest should be incorporated as refuges for wildlife and islands of native habitat. Some countries have legal requirements, or local FSC standards, which may define the minimum areas to be covered by this type of reserve. For example:
   - Brazil: legislation requires that 20 per cent of plantation areas should be set aside as natural forest reserves.
   - GB FSC Standard: 15 per cent of the forest area must be managed with biodiversity as a major objective including 1 per cent for long-term retentions and 1 per cent for natural reserves (in plantations) to be managed by minimum intervention.

3 Under FSC-based standards, plantations established in areas converted from natural forests after November 1994 are not normally eligible for certification. However, in circumstances where the forest owner or manager can provide adequate evidence that they were not responsible directly or indirectly for the conversion, certification may be allowed.

TABLE 8.5 Conservation and restoration of natural forest cover

FSC requirements:	Responsibility	Scale
1 Natural forests must not be converted to plantations	P/T	🌲
2 A proportion of the forest management unit should be managed for restoration to natural forest cover	P/T	🌱
3 Plantations established in areas converted from natural forests after November 1994 are not normally eligible for certification under the FSC scheme	P	🌲

Scale considerations! ♀

Restoration of a proportion of the forest operation to its natural forest cover is feasible only by medium or large organizations. In small areas the forest manager should aim to add diversity within the plantation by using native species where possible, managing individual old-growth trees, streamside buffer strips and wildlife corridors.

Conservation and restoration of natural forest cover – see also:

Section 15.3: Writing a Management Plan

Section 15.6: Developing a Conservation Strategy and the management of High Conservation Value Forests

Performance Requirements

This table summarizes the requirements described in Part Two.

The forest management team may use this table to visualize which issues need to be addressed, and to prioritize areas where improvements can be made. Note down on this table (or a photocopy of this table) which performance requirements have been met, and which requirements still need to be implemented at your forest organization. Use the table to allocate responsibilities within the management team.

Key

Responsibility: Who is responsible for carrying out the requirement?

Suggested responsibilities are:
P = policy/managerial
T = technical staff
S = specialist input needed

Scale: What scale impact is the organization? (see Table Part 2-1)

 = requirement applies at all scales

◯ = small-scale organizations may treat requirement differently

Performance Requirements	Responsibility	Scale	Comments
THE LEGAL AND POLICY FRAMEWORK			
4.1 Compliance with legislation and regulations			
FSC requirements:			
(All three requirements are not specifically required by ITTO, but they are implicit in the guidelines)			
1 Compliance with local and national regulations	P	🌲	
2 Compliance with applicable international agreements	P	🌲	
3 Payment of all charges, fees and royalties	P	🌲	
4.2 Tenure and use rights			
Shared FSC/ITTO requirements:			
1 Long-term legal rights to manage the forest resource	P	🌲	
2 Recognize and respect local communities' legal or customary rights	P	🌲	

4.3 Commitment to SFM

Shared FSC/ITTO requirements:

1   Reinvest part of the financial benefits
    from forest management in maintaining SFM   P

FSC requirements:

2   Demonstrate long-term commitment to SFM   P

SUSTAINED AND OPTIMAL PRODUCTION OF FOREST PRODUCTS

5.1 Management planning

Shared FSC/ITTO requirements:

1   Undertake management planning at
    appropriate levels   T

FSC requirements:

2   Periodically revise the management plan   T

3   Make a summary of the management
    plan publicly available   P

5.2 Sustained yield of forest products

Shared FSC/ITTO requirements:

1   Set harvest rates at sustainable levels   T

2   Collect data defining sustainable
    production levels   S

3   Adopt a reliable method of controlling yield
    (eg AAC). Where data are unreliable, set
    production levels conservatively   T

4   Maintain records of actual production
    levels of wood and non-wood products   T

5   Periodically revise yield (AAC) levels   S

6   Document and justify the choice of
    silvicultural system   T

7   Properly supervise all harvesting
    operations and silvicultural prescriptions   T

5.3 Monitoring

Shared FSC/ITTO requirements:

1   Monitor environmental, financial and
    social effects of operations   T/S

2   Carry out post-harvest assessments   T

FSC requirements:

3   Make a summary of monitoring
     information publicly available   P

5.4 Protection of the forest resource

Shared FSC/ITTO requirements:

1   Protect forest from illegal harvesting
     encroachment and activities that are
     incompatible with SFM   T

2   Control inappropriate hunting, fishing,
     trapping and collecting   T/P

3   Establish a fire management plan and
     warning systems for the forest organization   T

5.5 Economic and viability and optimizing benefits from the forest

Shared FSC/ITTO requirements:

1   Ensure forest management is economically
     viable, taking into account full environmental   T
     and social costs

2   Safeguard multiple benefits of forests during
     all operations   T

FSC requirements:

3   Encourage optimal use   P

4   Encourage local processing   P

PROTECTING THE ENVIRONMENT

6.1 Environmental and social impact assessment

Shared FSC/ITTO requirements:

1   An ESIA should be carried out prior to
     site-disturbing operations   T/S

2   ESIA results must be integrated into
     management operations   T

6.2 Conservation of biodiversity

Shared FSC/ITTO requirements:

1   Conserve diversity at genetic, species
     and ecosystem levels   T/P

2   Establish conservation zones and
     protected areas, including representative   T/P
     examples of existing ecosystems

3   Ensure safeguards exist to protect rare,
     threatened and endangered species and   T
     their habitats

FSC requirements:

4   Genetically modified organisms must
    not be used   P

5   Biological High Conservation Values must be
    maintained or enhanced   S

6   Forest conversion to plantation or non-forest
    land must not occur   P

6.3 Ecological sustainability

Shared FSC/ITTO requirements:

1   Maintain processes of forest regeneration,
    succession and natural cycles   T

2   Develop and implement guidelines for
    the identification and protection of   T
    sensitive soil and water resources

3   Prepare and implement written guidelines
    for road construction and use   T

4   Develop and implement reduced impact
    harvesting and extraction guidelines   T

6.4 Use of chemicals (also see 8.4)

Shared FSC/ITTO requirements:

1   Minimize chemical use and adopt
    integrated pest management   T/S

2   Implement procedures for handling,
    storage and disposal of chemicals   T

3   Provide adequate training and equipment
    for chemical use   T

4   Special restrictions must apply to
    chemical use in sensitive areas   T

FSC requirements:

5   Recognized dangerous and banned
    chemicals must not be used   T

6   Use of chemical pesticides must be clearly
    justified and recorded   T

7   Document, monitor, strictly control and
    minimize use of biological control agents T/S

6.5 Waste management

FSC requirements:

1   Dispose of all waste properly, off-site
    where appropriate   T

2   Minimize waste from harvesting   T

THE WELLBEING OF PEOPLE

7.1 Consultation and participation processes

Shared FSC/ITTO requirements

1 Maintain consultations and encourage participation by people and groups affected by forest operations   T

FSC requirements:

2 Employ appropriate mechanisms for resolving grievances and providing compensation for loss or damage to rights, property resources or livelihoods   P/T

7.2 Social impact assessment

Shared FSC/ITTO requirements:

1 Carry out social impact assessment and incorporate results into management planning   S

2 Take measures to avoid possible negative social impacts   T

7.3 Recognition of rights and culture

Shared FSC/ITTO requirements:

1 Recognize and uphold legal and customary rights of local and indigenous communities to control management on their lands   P

2 Identify and protect sites of special cultural, ecological, economic or spiritual significance to indigenous peoples   T

FSC requirements:

3 Compensate indigenous peoples for the application of their traditional knowledge   P

4 Social High Conservation Values must be maintained or enhanced   P

7.4 Relations with employees

Shared FSC/ITTO requirements:

1 Meet or exceed all applicable laws and/or regulations covering health and safety of employees and their families   T

2 Provide adequate training for all staff   T

FSC requirements:

3 Provide opportunities for employment and

training to communities in or near the forest operation	P		
4 Guarantee the rights of workers to organize and negotiate	P		
5 5 Comply with applicable conventions of the International Labour Organization	T		

7.5 Contribution to development

Shared FSC/ITTO requirements:

1 Contribute to an equitable distribution of the benefits, costs and incentives of forest management	P		

FSC requirements:

2 Strive to strengthen and diversify the local economy	P		

PLANTATIONS

8.1 Plantation planning

Shared FSC/ITTO requirements:

1 Consider impacts of plantation on local communities, biodiversity and landscape: design and layout to promote protection of natural forests	P/T		
2 Pay special attention to social issues of land acquisition for plantations	P		
3 Design layout to include a mosaic of stands	T		

ITTO requirements:

4 Carry out detailed site survey as the basis of planning	T		
5 Carry out integrated resource inventories to evaluate risks and opportunities of plantation	T		
6 Before final harvest, plan for following generation	T		

8.2 Species selection

Shared FSC/ITTO requirements:

1 Select planting stock based on appropriateness to the site conditions and management objectives	T		
2 Native species should be used in preference to exotics	T		

3	Diversity is preferred at all scales	T	
4	Carry out research to determine suitability to site	T/S	

8.3 Soil and site management

Shared FSC/ITTO requirements:

1	Minimize soil exposure and maintain or improve soil structure and fertility	T	
2	Establish completely protected streamside buffer zones along all watercourses	T	
3	Use low-impact weed control methods	T	
4	Restrict applications of fertilizer	T	

ITTO requirements:

5	Restrict intensive site preparation to suitable sites	T	

FSC requirements:

6	Maintain water quality, quantity and drainage patterns	T	

8.4 Pest and disease management

Shared FSC/ITTO requirements:

1	Carefully match species to site and cultural practices to prevent pest and disease outbreaks	T	
2	Avoid indiscriminate, blanket applications of chemicals; aim to reduce chemical application in general	T	
3	Use integrated pest management	S	

8.5 Conservation and restoration of natural forest cover

FSC requirements:

1	Natural forests must not be converted to plantations	P/T	
2	A proportion of the forest management unit should be managed for restoration to natural forest cover	P/T	
3	Plantations established in areas converted from natural forests after November 1994 are not normally eligible for certification under the FSC scheme	P	

Part Three

Using an Environmental
Management System

Introduction to Part Three

An Environmental Management System (EMS) provides a framework for achieving and maintaining high standards of environmental performance, such as those set out by the FSC Principles and Criteria and the ITTO Criteria and Guidelines.

An EMS can assist the forest organization to:

- plan, control and monitor its environmental performance;
- comply with legislation;
- meet required standards (eg FSC and ITTO);
- continually improve its environmental performance.

The EMS sets out the management tools which are generally accepted to lead to improved environmental and social performance. The EMS described here comprises the basic elements of policy, planning, implementation, monitoring and management review. The feedback of information from monitoring and management review into improved planning and on-going management should lead to the continual improvement in the environmental performance of the organization. This is one of the fundamental goals of an EMS.

Box Part 3-1   The EMS Model and related forest management activities

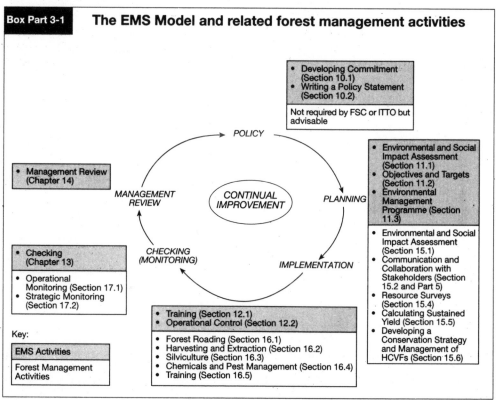

The EMS outlined here is based on the model used by ISO 14001 (see Section 3.3). The EMS model shown in Box Part 3-1 is not specific to forestry, and can be applied to any industry where improved environmental management is required. The forest management activities related to each element of the EMS are also shown. In Part Five, Box Part 5-1 shows how social issues are similarly linked with EMS elements and forest management activities.

Part Two examined the performance requirements of sustainable forest management, as defined by the FSC and ITTO. Part Four looks at the forest activities which need to be undertaken to meet these performance requirements. Part Three, using an EMS, provides a framework which may be used for organizing the activities outlined in Part Four. While parts or all of the EMS described here may be useful in meeting the requirements of SFM, it is not a requirement of the FSC or ITTO standards. The relationship between the performance requirements of SFM and the EMS approach outlined in this chapter is shown in Box Part 3-1.

Thus, the areas with which the forest manager is most familiar – planning, implementation and monitoring of forest operations – form the core activities of the EMS. The other management systems elements – policy, internal audit and management review – are intended to support forest management and encourage continuity of management, feedback and continual improvement.

The elements of the EMS covered here are:

 Chapter 10 Sustainable Forest Management Policy

 Chapter 11 Planning

 Chapter 12 Implementation

 Chapter 13 Checking

 Chapter 14 Management Review

10 Sustainable Forest Management Policy

A sustainable forest management policy is the public expression of the forestry organization's commitment to the goal of SFM. It provides an undertaking that forest management will be carried out in ways which are environmentally responsible, socially beneficial and economically viable.

The policy statement forms the basis of relationships between the forest organization and local communities, governments, buyers and other stakeholders. It lets stakeholders know what they can expect from the organization – and provides an opportunity for public discussion about operations. It should help build consensus and reduce conflicts about operations.

What is Required?

Two important aspects need to be addressed in developing a SFM policy:

1 developing commitment within the organization;
2 writing a public policy statement.

10.1 Developing Commitment

Implementing SFM entails many challenges and changes. It may mean significant changes in current practice, changes in lines of authority and the balance of power and may entail increased spending or lower revenues. Especially initially, these changes require energy and enthusiasm: none of this is possible without commitment to SFM.

Practical steps for developing commitment include:

- *Develop a* SFM *policy statement* setting out the common goals to which everyone in the organization should be committed (see below).
- *Publicly commit funds and staff to implementing SFM programmes*, including training in new skills and investment in environmental and social programmes.
- *Communicate and consult within the organization* to ensure everyone understands and participates in developing SFM systems.
- *Assure staff that appropriate training and support will be provided* to enable them to implement new working practices.
- *Seek advice and encourage participation* of people from outside the organization.
- *Explain benefits of improving forest management* and focus on people's hopes for the future (for example sustainable livelihoods, development of local businesses).
- *Be clear what changes are required*, when they will occur, who will be affected, and reassure the people affected by changes.

In large companies, management staff, and in particular the chief executive, need to be clear about their aims and personally committed to the vision of SFM. This will help generate commitment in the rest of the organization. The aim is for everyone within the organization to be committed to the vision, understand how SFM will be implemented, and what this will mean for their jobs, livelihoods and careers.

Box 10.1 | **Sample environmental policy – New Zealand Forest Managers (NZFM Ltd)**

New Zealand Forest Managers (NZFM) is committed to sustainable forest management. The Company will work with its business partners and other stakeholders to achieve continual improvement in our environmental performance.

NZFM will:

- Recognize the environmental requirements in management contracts, resource consents and in the Principles for Commercial Plantation Forest Management in New Zealand.

- Ensure that environmental management and performance is consistent with the Principles and Criteria of the Forest Stewardship Council.

- Ensure contractors and employees are committed to the Company's Environmental Policy.

- Continually review opportunities for further environmental improvements.

- Work in partnership with other business organizations, the Government and interest groups where opportunities exist to benefit both our business and the environment.

- Ensure this sustainable forest management commitment is integrated with other key business objectives of financial performance, operating efficiency, customer satisfaction, health and safety and good corporate citizenship.

Copies of our Policy are circulated to our business partners, including our contractors. Our Policy is available to the public from our office in Turangi.

Our employees are provided with copies of the policy, and receive information to help them understand its relevance and importance to our business.

Our senior managers will review our Policy annually. We will consult our business partners and stakeholders to ensure that our commitments are consistent with their requirements.

10.2 Writing a Policy Statement

The policy statement describes a broad set of economic, social and environmental objectives. It should outline how the organization intends to achieve these objectives, and what the benefits will be for employees, members of the organization and, in some cases, other stakeholders. The level of detail contained in the statement will depend on the size of the organization, and how much impact it may have on local communities, the environment and government.

The policy statement may be all that outsiders know about the forest organization. As a result the policy statement should be:

- committed to specific measurable standards;
- accurate, clear, concise and comprehensive;
- available to, and understood by, the whole organization;
- detailed enough for stakeholders to understand what the company is doing;
- short enough to be made easily available to the public;
- realistic enough to be achievable;
- dated and endorsed by senior management;
- reviewed and revised periodically as necessary.

Box 10.2 **Sample environmental objectives – NZFM Ltd**

Objectives 1 and 2 (of 5) are shown.

Environmental objectives are developed and reviewed annually, together with a description of how each will be achieved during the forthcoming year. These environmental objectives and targets are derived from:

- The Environmental Policy, with its commitments to sustainable forest management and improving environmental performance.
- The environmental requirements contained in the contracts, leases and plans for the forests that are managed by NZFM, RMA requirements, and applicable Codes of Practice which NZFM has committed to observe.

Objective 1

To manage and/or develop forestland for the benefit of the owners and their beneficiaries, ensuring forest utilization for the maximum practicable financial yield, consistent with forestry practices that ensure sustainable forest management.

This objective will be achieved through environmental and social effects assessment, and forest management and harvesting strategies and plans, developed in consultation with owners and lessees.

Parameters monitored
- Compliance with Leases, Management Contracts, and the Environmental and Management Plans of owners and lessees.
- The balance of planted and harvested land.
- Projections of future harvest areas.

Responsibility	*Authority*
Planning, Forestry Operations, and Logging and Distribution Managers	General Manager

Objective 2

To ensure forest management operations do not adversely affect Waahi tapu, historic, and other sites of cultural significance.

This objective will be achieved through forest management and harvesting strategies and plans, developed in consultation with forest owners and lessees. The location of and information about waahi tapu, historic and other sites of cultural significance remain the property of the owners, and that information will remain confidential.

Parameters monitored
- Compliance with Leases, Management Contracts, and the Environmental and Management Plans of owners and lessees.
- Evidence of disturbance to waahi tapu, historic and other sites of cultural significance, as revealed during inspections of operations.
- Feedback received from beneficial owners.

Responsibility	*Authority*
Planning, Forestry Operations, Logging and Distribution Managers	General Manager

Scale considerations! ♀

A policy statement should be detailed enough for stakeholders to understand what the organization intends to do, but it also needs to be short enough to make public availability practical. A large company, for example, may have a one-page summary of its policy for the purpose of public dissemination (see Box 10.1), and more detailed strategies (for instance covering labour relations, the environment, health and safety) to guide the company management and its employees (see Box 10.2). A public summary is sometimes published as an illustrated brochure for publicity.

For a small organization, a one-page policy statement might be sufficient.

Practical steps in writing a policy statement include:

- Nominate a lead person to guide the process.
- Define, with colleagues, the fundamental beliefs and vision of the organization.
- Consider the possible social, environmental and economic impacts of forest operations and outline how those impacts will be addressed (eg through an ESIA process – see Section 15.1).
- Consider commitments to external standards (legal requirements, management standards, eg FSC, ITTO).
- Consider broad objectives and targets of the organization and appropriate timescales.
- Set a schedule for review of the policy.
- Get the policy signed by senior management.

The policy statement should be widely available. This can be achieved by displaying a copy, for example as a poster, on a wall at the forest organization's offices, or making copies available to visitors at the reception desk. Copies could also be sent to people who enquire about the forest organization or forest operations, the policy may be proactively disseminated to stakeholders and the media, or it could be included in the annual report.

11 Planning

Planning in the context of an EMS provides the overall framework for carrying out the forestry-specific planning activities described in Chapter 15. The framework is intended to ensure that the strengths and weaknesses of the forest organization's current performance are identified and appropriate actions are defined to address them.

The EMS planning process together with the continual improvement feedback should assist the organization to adapt to changing circumstances. This might mean satisfying internal demands, such as priorities identified by the organization's own monitoring programme or ESIA, or conforming to outside pressures, such as legislative or market changes and research developments.

What is Required?

Planning for an EMS involves three main components:

1 Environmental and Social Impact Assessment
2 Objectives and Targets
3 Environmental Management Programme

11.1 Environmental and Social Impact Assessment

Environmental and Social Impact Assessment (ESIA) is an essential part of the EMS as it defines the positive and negative impacts that the forest organization is likely to have. Means of minimizing negative impacts and maximizing positive impacts are examined and results built into management through the rest of the planning and implementation process.

Because it is an essential forest management activity and a requirement of performance standards like FSC and ITTO, Environmental and Social Impact Assessment is discussed fully in Chapter 15.

11.2 Objectives and Targets

Objectives and targets are a statement of the specific performance levels the forest organization is aiming to achieve. They set out in a clear way the goals of the forest organization, the short-term targets to achieving those goals and the timescales within which they will be achieved.

Objectives and targets of the forest organization are usually based on:

- the vision and goals of the forest organization itself, defined in its policy statement;
- legal requirements which must be met;
- issues arising from the ESIA which must be addressed;
- gaps between current standards and the requirements of external standards such as FSC or ITTO which need to be met.

Objectives are broad statements of intent, often laid out in the company policy statement. These may include requirements from external performance standards to which the organization is committed, such as FSC or ITTO, and local legislative requirements.

Targets should be specific, quantitative and measurable, so that progress towards achieving them can be evaluated.

Some basic guidelines for setting objectives and targets:

- *Every department*, and, therefore, every member of staff, should be involved in achieving some of the objectives and targets.
- *Never set targets which are unachievable*. Sometimes targets may be too optimistic and need revision, but they should always be set with the intention of meeting them.
- *Review performance* regularly and be sure to acknowledge progress as well as questioning targets which have not been met.
- *Link objectives and targets* resulting from the ESIA with those defined as part of the management planning and conservation planning processes (see Sections 15.3 and 15.6).

11.3 Environmental Management Programme

The Environmental Management Programme provides the overall framework for achieving the objectives and targets. It sets out the specific actions necessary for getting from the current situation to the desired situation.

The Environmental Management Programme helps the organization adapt to changing circumstances.

Practical steps in developing an Environmental Management Programme include:

- Prioritize areas where improvements in performance are most urgently required.
- Set clear, measurable objectives and targets.
- Set milestones to be achieved by key dates.
- Define the financial, equipment and human resources necessary.
- Designate responsibility for carrying out each action.
- Examine current management practices for changes which need to be made in order to achieve the objectives.
- Ensure that adequate resources are allocated to enable the programme to be fulfilled.

The Environmental Management Programme might address any of the practical forestry activities covered in Parts Four or Five. For instance, preparing a conservation strategy (Section 15.6) might be a long-term objective of the forest organization, with specific targets of:

- assessment of the forest for High Conservation Values;
- definition of vegetation types in need of protection;
- definition of other areas where modified management prescriptions are needed (eg for protection of water sources);
- preparation of criteria for identifying areas to be protected;
- identification and survey of the protected areas;
- preparation of management prescriptions for each area;
- definition of monitoring and protection procedures.

For each of these targets the practical steps described above could be taken, in terms of definition of timescales, responsibilities and resources.

The Environmental Management Programme should aim to strike a balance between being sufficiently demanding to result in discernible progress, yet realistic enough to ensure that the objectives are not perceived as unachievable.

12 Implementation

Successful implementation of an EMS relies on planned activities being carried out effectively and consistently according to set rules or procedures, by appropriately trained personnel. Careful control of implementation of operations is essential, particularly in situations where staff turnover is relatively high and continuity of management may be lacking – as in many forest organizations.

What is Required?

Two major elements of implementation of the EMS are considered here:

1 training
2 operational control.

12.1 Training

Training is an essential component of any management system and SFM cannot be achieved without considerable investment in development of capacities and skills. A systematic approach to training is a fundamental element of the EMS: procedures for the identification of training needs, planning and implementation of appropriate training, and feedback on effectiveness of training should be part of any EMS.

Because it is an essential forest management activity and a requirement of performance standards like FSC and ITTO, training is discussed fully in Section 16.5.

12.2 Operational Control

Operational control defines the way that forest management should take place on a day-to-day basis to ensure high quality results. Documented procedures or appropriate guidelines should set out how each important activity should be carried out. As long as the procedures are followed, the results of the operation should be predictable and of a high standard. If documented procedures defining acceptable practices are not available, it is less likely that the desired results will be achieved. The important activities for which documented procedures need to be drawn up are generally identified as part of the ESIA process.

A procedure can be defined as any form of documented instruction, from a conventional written guideline to a numbered cartoon, for example:

- a comprehensive code of practice or set of guidelines for forest operations, such as the Reduced Impact Logging Guidelines for Indonesia produced by CIFOR; .
- a more specific guideline, for example the tree-marking rules prepared by the Malaysian–German Sustainable Forest Management Project in Sabah (see Figure 16.1);
- a cartoon, flow chart or diagram pinned up on a wall or notice board.

Box 12.1	**An example of an implementation procedure: Part of Tilhill's clearfelling procedure (UK)**

UPM

Tilhill
Environmental Management System

Site Planning, Sensitivities & Hazards

Thorough planning of harvesting and haulage allows environmental risks to be effectively controlled with minimum disruption to operations. Planning should commence as early as possible to take account of unknown site factors or road upgrade requirements.

Prior to any clear-felling operation beginning or being made available as a standing sale, the forest manager will identify and map the following features associated with the site and forest haulage routes.

A. THE SITE

i   The area to be felled. This should also be clearly identifiable on the ground.

ii   Any watercourses crossing the site, or any running drains crossing the site which discharge directly into watercourses. Any water bodies on or close to the site. Any drinking water supplies on or close to the site.

iii   Any public or permissive footpaths on or close to the site, any other known public use.

iv   Any designated sites or areas of importance for:
 – archaeology
 – flora
 – nest sites of raptors or other specially protected birds
 – badgers' setts
 – bat roosts
 – other protected species.
 These should be adequately identified on the ground and where necessary advice should be sought from the relevant conservation agency.

v   Trees to be retained as raptor perches or for deadwood habitat. Aim for 3/Ha.

vi   Site safety hazards including, overhead cables, buried cables and pipelines, old mine workings. These should be clearly identified on the ground.

vii   Areas of difficult working including known soft ground and steep slopes.

viii Preferred extraction routes taking account of all those factors identified above.

ix   Areas of flat hard ground suitable for machinery maintenance and fuel storage where spillages may be contained away from watercourses.

B. HAULAGE

i   Suitable forest roads to be used for haulage from site.

ii   Suitable areas to be used for timber stacking away from watercourses and major drains. Include hard standing areas for loading of flat bed trailers where required.

iii   Suitable turning areas for lorries.

iv   Hazards associated with the forest haulage route including, overhead cables, pipelines, watercourses and water bodies, bridges, weight and speed restrictions.

v   Sensitive neighbours, areas where livestock are unfenced, gates which must be kept shut, other road users.

vi   Weight restrictions and low bridges on the county road network. Where local authority timber haulage road agreements exist the designated county roads should also be identified.

vii   Large-scale clear-felling operations require considerable planning and often involve road upgrading, installation of lay-bys and turning circles. Careful planning of this infrastructure will dramatically improve the ease of operation and controlling of environmental hazards.

Issue No: 2	Clear-felling	Issue Date: Aug 2001	Issuer: SJ

The format for a procedure should take account of the level of training, skill and literacy of those people who will be expected to refer to it. Appropriate language must be used; for example, a document in the national language may be useless if the field staff can only read a local language. One way to ensure procedural guidelines are effective is to test them on a number of potential users before going ahead with the final version.

A procedure should define the requirements in a clear way, to ensure that the task can be carried out effectively. It does not have to document every detail about a job.

Practical steps for developing a procedure – consider:

- *Purpose* – why is the procedure necessary?
- *Scope* – what activities are covered?
- What are the required *standards*?
- What are the *methods* involved?
- What is the *sequence of events*?
- *Responsibility* – who does what (by job title)?
- *Decision-making* – who makes decisions and what authorizations are required?
- What *inspections* are required?
- What *records* are kept?
- *References* – other technical documents containing information necessary to carry out the procedure.
- *Definitions* – explanation of any unusual terms.
- *Documentation* – any standard forms or records that are generated by the procedure.

Documented procedures should be in place for all important forest management activities. Forestry activities for which procedures are considered especially important include road construction and use, harvesting and extraction, silviculture, chemical use and monitoring. Guidance for issues which should be considered when developing such procedures is provided in Chapter 16.

The same documented procedures should be central to training programmes (see Section 16.5), particularly for less experienced staff and in situations where staff turnover is relatively high.

Developing a set of procedures can provide an opportunity for a critical evaluation of current practices within the forest organization. This may reveal that some activities are being implemented differently by different forest managers, or that operations in a specific area are not as efficient as they could be. A sample procedure covering clear-felling operations is shown in Box 12.1.

13 Checking

Checking is a term used to include all monitoring and feedback activities. Information is collected about how effectively management prescriptions are being implemented and what effects they are having. This provides a feedback mechanism into future planning and management activities.

All forms of checking enable the forest manager to assess technical, environmental and social performance. Baseline information to inform planning may be collected through an ESIA, inventory and consultation (see Chapter 15). Further information showing the effectiveness of management and the changes in resource quality and quantity is built up by monitoring over time. Practices can be modified according to the data collected, ensuring that forest management continually adapts, improves and develops to meet the organization's overall aim of sustainability.

What is Required?

There are two types of checking required as part of an EMS:

- *Monitoring* – the recording of information to track performance. Monitoring can range from the informal collection of observations, through the use of basic checklists and procedures, to sophisticated and extensive surveys and experiments. It can be quantitative and/or qualitative. Monitoring is considered in Chapter 17.
- *Internal audit* – regular checks that specific elements of the EMS conform to plans and procedures, and are being properly implemented. Internal audit aims to ensure that all operations are carried out to required standards and to identify where corrective or remedial action is needed. This is relevant to all forest organizations, regardless of size or scale of impacts.

PLANNING AN INTERNAL AUDIT PROGRAMME

The following practical points should be considered when planning an internal audit programme:

- *Scope* – what operations (for example, harvesting) or which forest management units (for example, compartment or block) are covered.
- *Schedule* for different operations/areas.
- *Responsibilities* for auditing different operations.
- *Skills and training* of auditors.
- *Standard audit protocol* – the order of events, composition of audit teams, methodology for collecting evidence and recording of findings (such as the use of checklists).
- *Format* for audit reports.
- *Procedure* to feed back results and ensure that corrective actions are taken where problems have been identified.

WHO CARRIES OUT AUDITS?

As a means of checking and improving their own standards and providing feedback into improved management practices, all operators at all management levels should be encouraged to carry out audits of their own work. This can be a useful and cost-effective means of implementing improved standards where personnel are trained and well motivated. However, in order to provide more impartial checks on the quality of management, it is important that regular checks are carried out by auditors who are independent of the activity being audited.

Audits can be carried out by a person or department of the forest organization dedicated solely to this function; or audits can be undertaken by each department checking on another. For example, inventory staff might check the operations of harvesting teams, or roading managers check the silvicultural teams. The second approach has overall training benefits, and assists each section in understanding what the other sections are doing, but it needs careful management. It may be more efficient for large organizations to have dedicated auditing staff.

It is essential that audits are carried out in a consistent manner, especially if they are not all carried out by the same team. All auditors therefore need to be trained in use of the audit protocols. Performance of auditors themselves also needs to be periodically reviewed.

AUDIT REPORT AND CORRECTIVE ACTION

Each internal audit exercise should result in some form of audit report and specify corrective actions. The audit report should highlight aspects where planning, implementation or monitoring programmes do not comply with required procedures and standards. It should also specify the corrective action to be taken.

In general, if an operation has a high degree of risk, that is, if the activity has a significant chance of being performed inadequately, or if the consequences of inadequate performance would be severe, then auditing should be more frequent.

Internal audits provide feedback into the cycle of management planning, as an organization tries to achieve or maintain SFM. Within an EMS, the formal means of providing such overall feedback is through a management review (see Chapter 14).

14 Management Review

The management review provides an opportunity to look systematically at the effectiveness of the EMS. It evaluates performance and progress against policies, objectives and targets. For example, public policy statements, operational procedures, monitoring programmes, training programmes, consultation procedures and audit results may all be given an objective review.

A review of this sort is vital to the continual improvement of the forest organization's environmental performance. The management review assists the forest management team to identify which policies or strategies are working, which need to be improved and which should be abandoned. It also helps in communicating progress within the organization.

What is Required?

A management review may comprise a formal meeting of people with key forest management responsibilities. The meeting should run to a standard agenda, and brief reports should be prepared on each item for everyone at the meeting to consider. Ideally these reports should be prepared and circulated in advance of the meeting. Alternatively, a management review may be held over a period of time, looking at each aspect in turn.

The management review should consider all aspects of the EMS. Notes should be taken at each management review, and any agreed action should be recorded, along with responsibilities for action and a time schedule.

Issues which might be covered by management review include:

- results of internal audits and assessments by external bodies;
- the extent to which objectives and targets have been met;
- the suitability of the EMS and Environmental Management Programme;
- review of actual or forthcoming changes in legislation or of external standards, and their implications;

Box 14.1   Addressing common problems with EMS

There is too much EMS documentation which does not match actual practice.

1   Where possible use existing documentation systems as part of the EMS. For example, the management plan can be a central part of the EMS planning.

2   Ensure that everyone participates in developing the EMS (and especially work procedures) from the outset. If the documentation is being developed by one person (typically the system manager or external consultant) make sure they work closely with the people involved in implementing it.

3   When writing procedures, use one lead person or a small group to develop the format and a few sample procedures. Then encourage each department to write their own procedures. Discuss and amend with the lead person/group.

- concerns of stakeholders;
- technical overview of operations, including positive and negative aspects;
- actions to be taken on problem areas of operations;
- progress of improvement programmes;
- performance of contractors;
- training programmes;
- new developments;
- responsibilities and time schedules for further action.

Part Four

Meeting the
Requirements

Introduction to Part Four

Part Four describes *how* the requirements outlined in Part Two can be met. It is based around the concepts of planning, implementation and monitoring, which are basic parts of forest management as well as being integral parts of the Environmental Management System described in Part Three.

This part of the handbook describes forest activities which will be very familiar to many forest managers. A handbook such as this cannot aim to provide a technical manual on all areas of SFM activities: this information must be found locally to ensure that it is adapted to prevailing conditions and fits with local or national best practice. Part Four aims to provide common approaches to forest management activities, and particularly to explain how the specific requirements outlined in Part Two can be met.

Part Four contains a brief summary of issues surrounding collaboration and communication with stakeholders. This is further developed in Part Five, but has been included as part of Part Four because of the importance of working with other people at all points in forest management. Box Part 5-1 shows the ways in which social issues link in with the forest management activities described in Part Four.

The three main areas covered in this part are:

Chapter 15 Planning

Chapter 16 Implementation in the Forest

Chapter 17 Monitoring

15 Planning

The planning of forest activities includes all the preparatory work to guide implementation in the field. This section covers the large-scale planning involved at the level of the FMU as a whole: operational planning of individual operations (such as harvesting or road construction) is considered in Chapter 16. The planning of forest activities described in this section also links to planning in the context of an EMS, described in Chapter 11.

FMU planning should ideally be carried out prior to initiating forest management activities. However, few forest organizations are in the position of starting from scratch and FMU planning should be incorporated as soon as feasible.

Six areas of planning are covered although these should not be considered to be sequential steps:

1 environmental and social impact assessment;
2 communication and collaboration with stakeholders;
3 writing a management plan;
4 resource surveys;
5 calculating sustained yield;
6 developing a conservation strategy and management of High Conservation Value Forests.

15.1 Environmental and Social Impact Assessment

WHY IS EIA IMPORTANT?

Environmental and social impact assessment (ESIA) is a way of examining forest management in terms of the effects it has on both the environment and people. An ESIA is a powerful tool to help in decision-making and planning. It helps to clarify the choices to be made, highlighting the expected and actual consequences of a project or operation.

The concept of ESIA has grown out of the widespread use of environmental impact assessment (EIA), which originally put less stress on the social aspects of management. Social and environmental impacts often overlap and, therefore, consideration of social impacts has become more common.

The basic ethos of ESIA is to ensure that forest managers are aware of the potential consequences of their actions before they are implemented. It allows the forest manager to consider alternative courses of action and their probable impacts, and to make an informed choice between them. The emphasis is on the prevention of environmental and social problems rather than cure, identifying the potential impacts before they occur, mitigating the negatives and enhancing the positives.

The earlier potential or actual impacts of forest management are identified, the easier it is to avoid or mitigate them. It is generally easier to change plans than to alter projects that have already been implemented, and it is easier to change plans at an early stage of project design than in the final stages. For these reasons, it is vital to ensure that the ESIA approach is adopted throughout the forest organization.

ESIA can act as a guide to the forest organization in setting its objectives and targets (see Chapter 11) and provide a basis for the management planning process (see Section 15.3). The data collected for the assessment can also be used as a baseline, against which to monitor the actual impacts of forest operations in future (see Chapter 17).

A formal ESIA is commonly used in forestry organizations in two situations:

- evaluation of the potential impacts of planned activities in special situations, such as prior to a large afforestation project or new road construction;
- evaluation of the actual impacts of current forest management activities, feeding into forest management planning and the improvement of on-going activities.

At an operational level, an on-site evaluation of environmental and social impacts should also be a routine part of operational procedures. It is a useful way to incorporate a 'look-before-you-leap' approach into the daily routine of forest operations. This is not a full ESIA, but aims to provide a quick assessment of the possible impacts of individual operations. It may be carried out by a manager, supervisor, or responsible staff member. On-site assessments are discussed at the end of Section 15.1.

An ESIA may be a legal requirement for forestry operations of a certain size or type, depending on local legislation. If this is the case, the legislative requirements must be followed. Finally, some forest management standards (including FSC and ITTO-based standards) also require an assessment of environmental and social impacts, for which the ESIA methodology can provide a basis.

WHAT IS REQUIRED?

An ESIA is a method of identifying and considering the impacts of a proposed or actual management regime in an open and transparent manner. It involves an analysis of the physical, biological and social implications of the operation.

The exact requirements for how an ESIA should be carried out vary from country to country, but there is wide agreement on the key steps. A large number of ESIA methodologies have been developed and a wide range of guidance is available internationally (see Appendix 4.2 for details). Several international agencies (such as the World Bank, United Nations Development Programme and the UN Food and Agriculture Organization) have developed their own ESIA requirements. In some

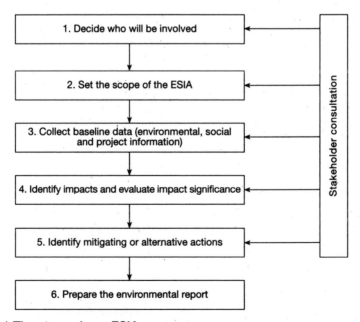

FIGURE 15.1 **The stages in an ESIA**

countries, ESIA (or EIA) is a legal requirement for large, potentially high-impact projects. National guidance exists in over 60 countries worldwide, several with specific guidance for forestry projects.

Although there is no single method for carrying out an ESIA, it normally consists of a number of steps, which may be tailored to fit the local situation. These steps are shown in Figure 15.1 and discussed below.

1 Decide who will be involved

It is unlikely that a single person will have a broad enough range of skills to carry out an ESIA alone, except for the smallest forest. A mix of technical, environmental and social skills are needed to cover all aspects of the ESIA, so a team effort is normally required. The forest organization needs to appoint a 'champion' to lead the process and coordinate the team.

An ESIA might be carried out internally by staff of the forest organization or by an external team or company. An internal assessment has the advantage of helping the whole organization to understand how they are currently performing and provides an opportunity to think about how to improve performance. However, the forest organization may not have the necessary skills and experience to carry out a full ESIA, and may need to bring in external specialists. Bringing in stakeholder and specialist inputs early on in the process can also save time and resources later in the ESIA by highlighting important issues and avoiding easily identified problems.

An external study may be carried out by natural or social scientists and specialists in ESIA and is likely to bring expertise and new perceptions to the assessment. An external study may be required by law, requested by investors or local regulators, or desired by the forest organization themselves to provide external views on their management.

Regardless of whether the ESIA is carried out internally or externally, it is important that:

- A specific member of the forest organization's management team is responsible for coordinating the process and maintains an overview of the assessment.
- A wide range of employees within the forest organization are involved in the assessment, to ensure that everyone understands the assessment and is committed to taking action based on its findings.
- Specialists are brought in if needed to provide expertise on particular topics. This is particularly likely to happen with an internal ESIA.

External stakeholder groups, such as local communities, government departments and environmental and social NGOs need to be involved throughout the ESIA. This involvement is likely to be repeated: consultation in the early stages of scoping may help identify which issues the ESIA should examine, while subsequent consultation might deal with what data should be collected, what impacts should be considered significant and what mitigating or alternative actions are likely to be acceptable. This consultation should build on, and contribute to, the process of identifying stakeholders and developing collaboration (see Section 15.2 and Chapter 19).

External stakeholders can be involved in the ESIA process in a number of ways, including:

- inviting stakeholder representatives to be a part of the assessment team, or to help with coordination;
- inviting stakeholder representatives to attend one or two assessment team meetings;
- holding a meeting between members of the assessment team and stakeholder groups or representatives;
- inviting stakeholder groups to review the team's findings in stages.

2 Set the scope of the ESIA

Almost every forestry operation will result in some environmental and social impacts. Setting the scope of an ESIA identifies the aspects of the forest operations that are likely to give rise to key

Box 15.1   **Sources of information which may help forest managers**

Legislation and regulations

All relevant legislation must be considered. This may include forestry acts, environmental law, employment law, laws or treaties relating to indigenous rights, highways or transport law, wildlife law and any other national or local laws or regulations likely to affect the forest operation.

International agreements

Usually, when a government signs an international agreement (such as the ILO Conventions or the Convention on Biological Diversity) it incorporates the requirements into national law. Sometimes, however, there is a delay in doing this, during which the requirements of the international agreement should be considered when deciding issues to be covered by an ESIA.

Requirements of standards

If the forest organization is planning to be certified against an international performance standard, such as FSC, then all the requirements of the standard must be considered. These are discussed in detail in Part Two.

Industry or Association Code of Practice

If the forest organization belongs to an industry association which has a set of guidelines or a compulsory standard, then the requirements of the association must be considered.

Views of stakeholders

It is very important to consider the needs of external stakeholders. For example, local communities may have very strong feelings about issues such as fuelwood, sacred sites, water pollution, employment and so on. The best way to find out what these local requirements are is to talk to people.

Internal ideas & experience

Internal ideas and experience are a vital part of the process. Not only do members of staff have the deepest knowledge of many of the issues which need to be discussed, particularly those involved in field operations, but it is also an excellent way to ensure that everyone feels involved in implementing the results of the assessment. Existing forestry records may have considerable information about actual and past impacts of forest management.

---

issues and makes sure that the assessment is focused on these. This should allow the ESIA process to discount other aspects that are not important. A good scoping process aims to:

- decide which issues need to be investigated by the ESIA and which ones do not;
- decide how baseline data should be collected;
- identify suitable methods for evaluating or predicting environmental and social impacts; and
- identify suitable methods for deciding whether the impacts are significant.

It is important to get the scoping exercise right because setting the wrong scope wastes time and money by collecting unnecessary information, investigating non-significant impacts and, perhaps, missing significant impacts.

Some sources of information that can help to identify the key issues for consideration during the ESIA are shown in Box 15.1. Effective scoping also needs to incorporate good consultation with stakeholders. The process should begin by identifying individuals, local authorities and interest groups, experts, NGOs and communities who may be affected by the forest operations.

Although the issues raised will vary to some extent, most environmental and social impact assessments of forest organizations are likely to cover some or all of those issues listed in Box 15.2. During scoping, these need to be considered to decide whether they should be investigated further.

The scoping process should result in the identification of a list of key issues that will be comprehensively addressed by the assessment. These should be the elements of the environment or society that are likely to be most impacted by the forest operation. At the end of the scoping process, the team should be able to say which issues are important and why, while also recording the issues which need no further assessment and the reason for this decision.

In addition, the scoping stage provides an opportunity to discuss and agree the way in which the rest of the assessment will be carried out. Stakeholder and specialist input may help greatly in deciding how to assess and analyse some of the issues which have been identified through scoping.

3 Collect baseline data

The ESIA aims to assess the negative impacts that an operation or management regime is likely to have and what should be done to mitigate or avoid them, while building on the positive impacts. Collecting baseline environmental and socio-economic data is necessary in order to understand the current situation and evaluate potential impacts. These data also provide a baseline against which to monitor changes in future. It is an important step when using ESIA either to evaluate impacts of on-going activities, or potential impacts of proposed operations.

The scoping exercise is critical to ensuring that the collection of baseline data is focused and useful. Scoping should have highlighted the elements of the environment (physical, biological and social) that are likely to be affected by management. Where scoping has identified key issues that need to be investigated further, this stage provides an opportunity to collect information about the issues, to allow possible impacts and their significance to be evaluated.

Baseline data may need to be collected on some or all of the elements shown in Box 15.2. However, it is important to remember that data may not need to be collected for all the issues listed, but should be restricted to those identified as important during scoping.

Elements of the environment that are not likely to be affected by management really do not need to be investigated – attention, resources and time should be focused on the important elements. This is perhaps the most important rule of carrying out baseline studies. Two questions which should always be asked before collecting baseline data are:

- Why is this information required?
- What specific question will it help to answer?

Only if the information will answer an important question about the actual or potential impacts of forest management or the particular operation being considered, should it be collected.

There is a wide range of different types of data and collection methods that may be used at this stage. A common problem with ESIAs is that data collection often focuses on quantitative data (such as indicators of water quality or quantity) because it is easier to collect and analyse than qualitative information, such as people's perceptions or landscape values. However, it is important that qualitative information is included in the assessment and given full importance in the analysis, even if it is more difficult to capture. Different data may also be needed to demonstrate point information (ie what is the status of an indicator, such as the number of individuals of a particular bird species, at a particular time) and trend information (ie how has that number changed over time). It may be important to know this type of trend: for example, if the number of birds is declining over time, what will be the impact of forest management on the rate of decline? (See Appendix 4.2 for further reading on ESIA methodologies, providing more detail of data collection methodologies.)

Baseline information can be gathered from a wide range of primary and secondary sources. Government departments, non-governmental organizations, research institutions, forest workers,

Box 15.2	Issues on which baseline data may need to be collected during ESIA

Environmental values

- yields of timber and non-timber forest products;

- soil and water resources (quantity and quality);

- High Conservation Values, known or suspected to exist in the FMU;

- species diversity, especially rare or endangered species;

- ecosystem diversity, particularly less common forest types;

- carbon storage;

- waste disposal, minimization or recycling facilities;

- landscape values.

Social values

- access and use rights to the forest (tenure and control);

- economic livelihoods of local people (agriculture, paid employment);

- subsistence activities (fuelwood, wild foods);

- cultural and religious values (sacred sites);

- working conditions (forest workers, service providers);

- health and education facilities;

- recreational activities.

Other areas to consider

- performance of contractors;

- performance of suppliers;

- environmental and social aspects of emergency planning.

local people and private companies may all be able to provide useful data. Primary data collection should only be carried out where no alternative source of information is available.

At this stage, it is also common to collect together details of the forest organization itself (project information) that are likely to be associated with key issues identified in the ESIA. This is particularly important when the ESIA is evaluating the potential impacts of a proposed new project or operation. This includes information about the proposed or existing activities, scale and methods of operation and any major alternatives available, such as different sites or methods of operation. Information should also be collected on the 'without project' scenario: what will happen to the forest if the project does not go ahead? The consideration of alternatives will allow the analysis and comparison of the impacts created by different courses of action.

4 Identify impacts and evaluate their significance

Impacts of forest management may be positive or negative, long or short term, and permanent or temporary. The identification of impacts is an extension of the scoping and baseline data collection processes: having identified what are likely to be the major issues in the assessment, and collected baseline data to aid an understanding of the current situation, the next step is to identify existing impacts and predict future impacts of management. This needs to take into account the impacts of different management alternatives, as well as considering the situation in the absence of forest management. The following questions can be helpful in identifying impacts:

- What changes might occur in the area if forest management does not take place?
- What changes will occur due to forest management?

The impacts that are being considered at this stage must be very clearly defined in order to be meaningful. The indicators used to define them should have been agreed during the scoping process. For example, it is not helpful to ask 'what will be the impact of forest management on the water catchment?' because it is too broad to allow meaningful analysis. The impact might be on the quantity or quality of water, dissolved oxygen, suspended solids or a number of other characteristics, while 'forest management' might include all forest operations, harvesting certain blocks, road building and so on.

At this stage it is important to attempt to predict precise outcomes of specific management alternatives: for example, 'if we afforest 500 hectares in this water catchment with *Pinus elliotti*, leaving buffer zones of 20 m each side of permanent water courses, what will happen to the breeding population of otters in the catchment?' The answer to this might then be compared to other alternatives, such as no afforestation with the land used for agriculture, or afforestation with another tree species.

It is not always easy to identify or predict the likely impacts of an operation or forest management regime. The ESIA team is likely to need to use a range of methods to predict impacts, including:

- past experience, professional judgement and knowledge of similar situations;
- models, simulations or calculations;
- experiments or tests;
- historical and current maps;
- aerial photographs and other imagery.

The methods used to predict impacts should have been agreed by the ESIA team during the scoping process. Whatever method is used, it is important that it is made explicitly clear in the environmental report.

Once the range of possible impacts has been identified, they need to be prioritized according to their significance. This is perhaps the most difficult part of an impact assessment. To decide which impacts are significant, it is important to agree openly a 'threshold of significance' (or 'threshold of concern'). This is the point at which a particular impact should be considered significant and alternative or mitigating action should be planned and taken.

For some indicators, this threshold of significance may be quite simple to establish: for example, legal or international standards for water quality may exist, providing a clear threshold. For other impacts, the threshold of significance will not be clearly set and may not be quantifiable. The level of impact which people find acceptable may depend on their priorities and values. Stakeholder input and consultation with specialists is essential in arriving at an acceptable threshold of significance in this situation.

There are many methods for deciding which impacts are likely to be most significant. However, three priority questions may usefully be asked first of all:

1 Are we breaking the law? There may be legally set thresholds of significance for particular impacts. If any operations do not meet legal requirements it is important to address the issue immediately.
2 Are there issues that are particularly important to any key stakeholder group? Some impacts may be particularly problematic for some stakeholders. For example, if a village is dependent on a river for drinking water, then any pollution of the river, even at a low level, may be a serious problem.
3 Have we had serious problems recently? If adverse impacts have occurred, either environmental or social, these should be highlighted during the review process and addressed.

These three questions should identify, albeit not very objectively, the key impacts to be addressed immediately. More objective and systematic methods for evaluating potential significance can then be used to prioritize action on the remaining impacts that have been identified. These methods include matrices, scoring systems and ways of combining the risk and magnitude of possible impacts in deciding on their significance and therefore prioritizing them for attention. Appendix 4.2 shows further reading on ESIA, including techniques for predicting and evaluating impact significance.

Prioritization can be done for the whole forest organization or by operation. For small organizations, it is usually easier to prioritize for the whole enterprise. For medium and large companies, it is often useful to have a global set of priority areas, but then to also allow individual departments to decide if there are additional priorities for their operations.

5 Identify mitigating or alternative actions

The final stage of an ESIA is to plan how to address the impacts identified as significant. Bearing in mind that the impacts identified will often affect other people, it is important to involve those people in the development of the plan. This could be a continuation of the consultation process which identified the impacts in the first place.

Two alternative strategies can be considered in addressing impacts:

- *Mitigating actions* may need to be developed to reduce any negative impacts that have been identified (ie 'we want to do this, but because the impacts will be unacceptable, we will have to take extra (mitigating) measures in order to reduce them').
- *Alternative actions* may be necessary where impacts are potentially severe and cannot be mitigated, or where alternative ways of achieving the objectives are more effective (ie 'we wanted to do this, but accept that because of the impacts we will have to reach our objectives by doing something else'). Where the ESIA has been carried out before project implementation, alternative actions may be incorporated into a revised project design.

There may be occasions when it is impossible to avoid impacts entirely through mitigating or alternative actions. It may be appropriate in such circumstances to agree compensation with those affected by the forest management impacts, either financially or through services and facilities.

| Current situation | Mitigating actions | Alternative actions |

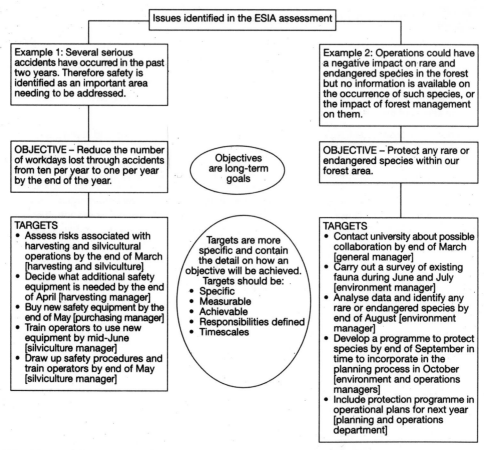

FIGURE 15.2 **Two examples of social and environmental objectives and corresponding targets**

Once any mitigating and alternative actions to be taken have been identified, an effective way to put them into practice is to set objectives and targets (see Section 11.2). Examples of objectives and targets are shown in Figure 15.2.

6 Prepare the environmental report

The ESIA process normally results in the production of an environmental report. There may be legal requirements relating to the content of the report and who it should be made available to. Where there are legal requirements, these need to be adhered to.

The environmental report usually has three main functions:

- As a record of the ESIA process, detailing the methods used and the findings of the assessment.
- Providing information about the actual and potential impacts of management operations and identifying mitigating and alternative actions. The information may be used by the forest management to inform their decisions, by government or regulatory authorities, or by other stakeholders.
- As a possible basis for consultation and negotiation with stakeholders. This will depend on the report, or a clear summary, being made available to stakeholders.

Box 15.3	**Addressing common problems with ESIA**

The ESIA results and recommendations are often not incorporated into forest operations, especially with ESIAs carried out by external specialists. Consider trying the following:

1 Make the ESIA ethos a part of the day-to-day routine. Ensure that staff automatically consider the effects of their actions before acting.

2 Prior to the assessment, agree clear terms of reference for the ESIA team and revisit them during the assessment if differences in interpretation occur. These should ensure that technical, environmental and social impacts are all addressed. It is likely that social aspects will be the most unfamiliar for the ESIA team, so make it clear that they must be considered.

3 During the ESIA, ensure that an appropriate, senior member of the forest organization accompanies the ESIA team at all times, to discuss findings as they occur. Try to ensure that other members of staff also work with the specialists during the assessment, improving their understanding of the results.

4 Try to ensure forest organization staff do not feel the ESIA has been forced on them as an additional constraint. Make sure that everyone understands that the ESIA should improve operations.

5 Ensure that a good ESIA team is provided. Some countries have professional associations that can provide contact details of experienced individuals and companies with specialist ESIA staff. Assessment teams need:

- a diversity of perspectives, preferably including more that three people (for a large organization), men and women;

- a mix of expertise and backgrounds including social, environmental and technical experience;

- an experienced team leader and team members who understand local and other stakeholder groups.

6 Require the ESIA team to produce a clear, concise report in appropriate language, with a summary in non-technical language. The focus should be on recommendations for mitigating and alternative action, NOT a long description of the ESIA process.

There is a danger with ESIA reports, especially those produced by external, specialist teams, that the environmental report is not fully utilized by the forest management. This may be because the report is long, technical, dull and difficult to understand. It is important, therefore, to demand from the start that the environmental report should include a clear summary of the most important actual and potential impacts and a discussion of how they can be mitigated or avoided.

· Although there may be legal requirements for the report format, the ESIA team should be able to produce a summary in easy-to-understand language, rather than technical jargon. The ESIA report must fully take into account the practical realities of forest management and therefore provide a basis for making genuine improvements.

Finally, the environmental report is there to be used and incorporated into management – it should not be put on a shelf and forgotten!

Scale considerations! ℗

Medium to high impact forest organizations are likely to need a formal ESIA carried out by a team of external specialists, so that adequate expertise is available to identify environmental and social impacts. Small-scale organizations may be able to carry out their own ESIA following the steps outlined above. This may be done over a period of time, with a focus on consulting with neighbours and local stakeholders who can help identify impacts. Small-scale organizations are unlikely to be able to carry out primary data collection or sophisticated evaluation of impact significance, and a common-sense approach should be pursued. A short social and environmental report, detailing the issues considered, people consulted and any actions to be taken as a consequence of the assessment, should provide a record of the process. On-site assessments of operational impacts (see below) are likely to be more important for small-scale organizations, than an overall ESIA.

ON-SITE ASSESSMENTS CARRIED OUT BY OPERATIONAL STAFF

An on-site assessment is useful for achieving a high standard of forest management because it enables staff to identify and minimize adverse impacts of their operations. It complements but does not remove the need for an in-depth ESIA.

Carrying out an informal on-site assessment ensures that staff are aware of the range of potential impacts, just by spending a short time analysing which negative impacts are likely to occur, and how these can be minimized before operations begin.

Operational procedures should provide guidelines for on-site assessments for all operations which potentially have significant environmental or social impacts. A brief checklist of issues to assess before starting work will help ensure that aspects are not forgotten or ignored. Often this is documented in the form of a sketch of a harvest site, or brief notes in a field note book. For particularly important operations, the standard of on-site assessments being carried out by operational staff should be checked regularly by supervisors, perhaps as part of the internal audit process (see Chapter 13).

An on-site assessment may be of particular value in order to meet specific objectives and targets identified as high priority by an in-depth ESIA. For example (Figure 15.2 example 1) an on-site assessment might be part of the safety procedures to be carried out before beginning harvesting operations on a new site. Once operators are trained (Target 4) they should be able to carry out on-site assessments of safety hazards.

A similar type of on-going assessment can be used for external social impacts. Whenever staff meet with representatives of stakeholder groups they should record any comments and discuss any operations likely to have an impact on stakeholders. This can be relatively informal, though it should be documented, if possible, in a field book.

See Section 15.2 and Part Five for more on stakeholder involvement.

Box 15.4 Addressing common problems with on-site assessments

Field staff do not carry out on-site assessments prior to site-disturbing operations. Undocumented on-site assessments are carried out initially, but become less consistently applied over time.
  Consider trying the following:

1   Provide quick and easy-to-follow procedures or checklists for on-site assessments.

2   Ensure the amount of writing required of field staff is kept to a minimum.

3   Ensure responsible field staff must sign off field books or field sheets to demonstrate on-site assessment has taken place.

15.2 Communication and Collaboration with Stakeholders

WHY IS COLLABORATION IMPORTANT?

One of the biggest changes in thinking about good forest management has been the realization that many people have a legitimate interest in the way a forest is managed. Forest management is no longer thought of as the exclusive job of the forest manager. The concept of 'stakeholders' has emerged in recent years: stakeholders are normally described as people who are interested in, or affected by, forest management. To ensure social sustainability, collaboration between managers and interested and affected parties in the process of forest management, is now recognized as vital (see Chapter 19 for more detail on working with stakeholders).

To ensure that the forest provides the widest range of goods and services possible, the forest organization must be aware of external stakeholders' expectations, needs and capabilities. This requires a process of on-going communication.

Many people are dependent on forests for specific products or services. This ranges from the almost complete dependence of forest-dwelling indigenous people to occasional recreational use by urban populations. Not all stakeholders want or need the same goods and services from the forest and there may be conflicts between different demands and different stakeholders.

Meaningful collaboration involves finding ways stakeholders can work with each other as well as with the forest manager, in order to find compromises and new solutions. Consultation with, and participation of, stakeholders in a forest operation can bring considerable benefits to the organization as well as the stakeholders (see Chapter 18). It should lead to improved long-term forest management and greater sustainability as well as reducing conflicts and criticism.

It can also ensure that the forest organization and the stakeholders are clear about each other's long-term needs and wants, which can help improve the sustainability of all operations. In reality, the scope of communication and collaboration in which a forest manager can engage may be restricted by bureaucracy within the forest organization and politics or poor forest governance structures outside the organization. A constant need to appeal to higher authorities in the forest organization takes time and may reduce the forest manager's stature and ability to negotiate. The forest manager needs to be aware of the limits to their scope for compromise with stakeholders and decision-making.

Where possible, the forest manager should try and ensure they have explicit support and authorization from senior management prior to starting a process of stakeholder communication and collaboration. This may be demonstrated, for example, by specific commitments made in the company policy, or objectives and targets (see Chapter 10).

WHAT IS REQUIRED?

As stakeholders are identified and as ESIA proceeds (see Section 15.1) the forest organization needs to encourage forms of participation which will develop further into good working relationships, in partnership with the key stakeholders. Identifying and working with stakeholders is discussed in detail in Part Five. This section provides a brief overview of some of the practical tools which can be used by forest management in developing a relationship with stakeholders.

Two main areas are covered:

1 identifying stakeholders;
2 developing and maintaining meaningful collaboration.

1 Identifying stakeholders

The first step in developing good communication and collaboration is to identify the important stakeholders. Some ways to do this include:

- identification by forest organization employees;
- self-selection;
- identification by other knowledgeable individuals;
- identification through written records, and population data;
- identification and verification by other stakeholders.

This is considered in more detail in Section 19.1.

2 Developing and maintaining meaningful collaboration

Collaboration between forest enterprises and stakeholders is still new, and everyone is still learning. Every situation will be different and it is important that all those involved, both the forest organization and stakeholder representatives, recognize the need to learn, to adapt and to improve.

The way in which collaboration develops is very dependent on the local context. The stakeholders themselves will shape the type of collaboration which is possible and desirable. For instance, factors which will have an influence include:

- the level of organization, education and access to resources which stakeholders have and the differences between stakeholders in these respects;
- the relations between the different stakeholders;
- practical issues of access to means of communication and transport;
- opportunity costs of collaboration for different stakeholders and groups.

Although some forest managers will be starting from scratch, most will have inherited existing relationships with stakeholders. This may include a history of mistrust and poor collaboration. Building genuine collaboration with stakeholders is not easy and needs to be worked at. Box 15.6

FIGURE 15.3 **The forest manager may want to communicate with a variety of stakeholders**

describes some elements of a strategy for collaboration. In developing such a strategy it may be helpful to focus on the following areas:

a) *Recognize community rules, local leaders and practices.* Recognition and promotion of local rights, including customary rights to land and trees; and recognition of community structures, leaders and practices are all important for establishing trust with local people, as well as working efficiently with them.

b) *Ensure communication, information flow and dialogue.* Ideally dialogue must continue beyond initial fact-finding so that key stakeholders are part of all major decisions on forest management design and implementation. It may take time to develop this level of trust and collaboration between the forest organization and stakeholders, but even if the process is faltering it is essential to keep communication channels open. Ideally:

  (i) **Provision of good information** at an early stage, and throughout the period of forest management, is essential to develop the high level of trust needed between stakeholders.

  (ii) **Do not restrict collaboration** or sideline key areas of activity or stakeholder groups because there is concern that they could reveal negative impacts, difficult situations or injustice.

  (iii) **Work with all stakeholders** even if they are difficult to involve (eg disempowered groups, women, inaccessible and distant communities).

  (iv) **Make sure that stakeholders are listened to with an open mind** and efforts are made to understand their points of view.

  (v) **Keep a record** of the processes that are followed (eg in writing, maps or video), those who are involved and the decisions which are made (but be sensitive to people who do not wish their names to be taken or their inputs quoted).

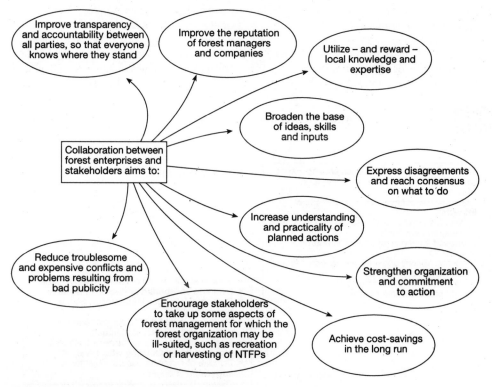

FIGURE 15.4 **The aims of collaboration**

Bear in mind that there are opportunity costs of collaboration for stakeholders

(vi) **Allow time** for agreements to be reached. This may mean allowing time for stakeholders' representatives to consult with their own constituencies.

c) *Form a stakeholder working group or liaison committee*. This should build directly on the results of the exercise of identifying who to work with – by inviting representation of these stakeholders onto the working group. The group should include company managers, workers and local people's representatives as well as the other key stakeholders identified. The group's general aim will be to develop mutual understanding and partnership through regular face-to-face exchange of information.

d) *Increase community capacity and control*. The use of participatory structures and methods needs to be combined with measures to support the politically weak groups who may have potential contributions to make towards SFM. This might include appropriate training and capacity-building for local people; design of farm forestry or other participatory schemes in line with agricultural activities; market assistance for small-scale producers, and establishment of outgrower schemes.

Box 15.5 **Typical information flows in collaboration**

- Aims and objectives of management and reason for seeking collaboration, including legal obligations;
- Expectations of environmental and social impacts;
- Types of issues to be addressed and what is considered negotiable;
- Proposed [changes in] approach and/or operations;
- Proposed options and roles and responsibilities of different stakeholders;
- Proposed forms of collaboration and cost-benefit sharing where relevant.

FOREST MANAGER

- Priority and intensity of feelings about issues;
- Desire to collaborate or degree of hostility;
- Proposed [changes in] approach and/or operations;
- Perceived impacts of forest management actions and suggested changes;
- Reactions to proposed forms of collaboration and cost-benefit sharing where relevant;
- Critique of forest managers' objectives and suggested changes.

KEY STAKEHOLDERS

| Box 15.6 | **Some elements of a strategy for collaboration** |

Summary of identification of main stakeholders, environmental and social impacts and issues arising/likely to arise in the future. Identify concerns, interests and positions as well as preferred priority. Clarify nature of representation and accountability.

Summary of initial visits, consultations and information exchanged. With whom were initial discussions held, what were main conclusions and what further discussions are needed? (Give names (where appropriate), dates and circumstances.)

Assess strengths and weaknesses of the collaborative approach. Consider resources, experience, credibility, legal/policy backing etc. What can be done to overcome weaknesses? Consider outside guidance, contracted specialists, assistance from others, and so on.

Propose people to participate in working group, liaison committee or local board. Give names and reasons for the people recommended for each type of meeting/grouping. Who will facilitate or chair the group? What will be the nature/mechanisms of the relationship with the forest managers? What is the strategy for evolving an agenda and broadening membership?

Schedule of actions. Give names of those responsible and time frames, then:

- appoint those responsible for steering the collaborative approach;
- hold meetings with own staff to gain commitment to process, and with stakeholders;
- train staff in consultation and collaboration approaches;
- prepare presentation of own position at meetings;
- prepare and conduct workshops, meetings with working group, liaison committee and so on;
- summarize and disseminate results of above collaborative events;
- implement agreed actions;
- monitor, review (on an annual basis at minimum) and further negotiate;
- document all the above.

Recording the process and the results. Keep attendance records, appoint people to take notes, draw maps, use tape recorders, video, etc. Documentation must meet legal requirements and will be needed if certification is sought.

Budgeting. Itemized budgets including travel, communications, meeting venues, consultant's services, report production and distribution, and so on.

e) *Plan direct provision of services.* Provision of basic needs and services for stakeholders including sustained employment, together with opportunities to develop skills. Support could also be provided through a transparent mechanism such as a committee with a defined annual budget. This committee should attempt to minimize the development of a culture of dependence or unrealistic expectations, or the delivery of services desired only by a minority.

f) *Commit to long-term accountability.* In an ideal situation, collaboration might include accountability of the forest organization over the long term, and agreements on a balance of corporate and local control which integrates commercial needs with locally perceived values of forests. This balance may be elusive at first but is a state which forest managers should be aiming to reach by continuous improvement in communication and collaboration with stakeholders.

Who needs a written management plan? I've got it all in my head...

15.3 Writing a Management Plan

WHY IS A MANAGEMENT PLAN IMPORTANT?

A management plan is an indispensable tool for planning sustainable forest management operations. It states objectives clearly so that everyone involved in managing the forest is working towards the same end.

A management plan should be set out in a systematic way and should include:

- the objectives of management;
- a description of the resources available;
- the proposed methods of achieving the objectives.

Sustainable forest management may mean changes to the way operations are carried out. Writing a management plan assists forest managers in moving from short-term crisis management towards long-term sustainable forest management. For example, including social objectives in forest management planning should result in fewer conflicts with stakeholders (and therefore less resultant crisis management) in the longer term.

A management plan provides the basis for continuity of management. If all plans and records are kept in the head of one forest manager, who then leaves, planning has to start again from scratch. A management plan makes the basis of future plans and activities clear and available to all. In addition a management plan provides a yardstick for monitoring performance, by forest managers themselves, or by governments and certification bodies.

The process of writing or revising a management plan provides a useful opportunity for the forest manager and other staff to:

- examine the current operation and clarify objectives;
- identify activities which are costing money but not providing significant benefits;
- increase resources available to activities which are providing benefits;
- take on board stakeholders' views (see Section 15.2);
- identify activities which are not compatible with the objectives of management.

Every forest organization's management plan will be different because it is written for a specific area of forest. Specified forestry practices in the management plan should be appropriate for the location. Once written, a management plan is not a static and unchangeable prescription: it should be revised periodically to take account of changing situations, technology, objectives and methods.

WHAT IS REQUIRED?

In some countries a management plan is required by law and must be submitted to the appropriate authority for approval. When this is the case, the plan should meet all legal requirements as well as those of the forest organization. In other countries, a formal plan is not required. Forest departments sometimes provide guidance on management planning even if it is not a legal requirement.

Writing a management plan can appear to be a daunting task, but in most cases much of the information required is already available, either in existing documents or in the heads of forest staff.

Five useful tips when writing a management plan:

1   Use a team to write the plan rather than giving all the responsibility to one person. However, one person will need to take overall responsibility for coordinating the writing and ensuring the plan fits together.
2   Approach the task step by step – the process of writing the management plan is as important as the product. Once draft sections of the plan are written you may need to revisit them in the light of other sections. This is an important and useful part of the process, not a waste of time!
3   Develop a working document which acknowledges difficulties as well as opportunities, and makes feasible plans for dealing with them; don't try to produce a glossy brochure which just makes everything look nice.
4   Plan realistic and achievable future activities to ensure that the management plan can be implemented.
5   Keep it short and succinct – include only relevant information and cross-reference to other sources of information (eg ESIA reports).

DEVELOPING A MANAGEMENT PLAN

The management planning documentation draws together and builds on information about the forest, management objectives, planned activities, resources and constraints in order to form a good basis for forest management. It provides an opportunity to combine sources of information and plans from diverse areas such as Environmental and Social Impact Assessment and timber inventories. It also provides a basis for discussion with stakeholders. The management plan must be useful for the forest manager: if it is not useful it will sit on a shelf gathering dust.

Many forest managers already have a management plan of some sort which can form a useful starting point. If the plan is being updated in order to meet new legislation, or a forest management standard, it is essential to know what specific additional information it needs to contain.

The following guidance may be useful in developing, revising or improving a management plan:

1   Start by collecting together all the available sources of information on which the plan will be based. This is likely to include:
    • inventory information;
    • results of Environmental and Social Impact Assessment;
    • results of any assessment of High Conservation Values in the FMU and conservation planning;
    • outcomes of any stakeholder consultation or collaboration;
    • research on silviculture, growth and yield, and harvesting impacts;
    • results of any monitoring.
2   As a preliminary step, it may be useful to discuss and try to agree management policy, objectives and a crude outline of work programmes with stakeholder representatives, as well as senior

---

Box 15.7 **An example of forest management plan objectives from Budongo Forest Reserve, Uganda**

The objectives of management are to:

- conserve 'in-situ' forest biodiversity and ecological conditions;

- produce economically the optimum sustainable hardwood timber of high value especially the mahoganies;

- integrate the communities surrounding Budongo Forest Reserve into management through Community Forest Management;

- promote and develop recreational facilities for the people of Uganda and others;

- carry out research in order to obtain information on various aspects of tropical high forest dynamics.

---

managers and employees of the forest organization. Once the objectives and programmes are better defined, they will provide the framework for the detailed management plan. It is therefore important that a senior management representative of the forest organization authorizes them and the management plan itself.

3 Set clear, appropriate objectives for management of the FMU, which are consistent with the resources available. Some flexibility may be necessary to allow the forest manager to respond to changes in markets, legislation or environmental or social conditions, but the main objectives of management should be clearly defined (for example, see Box 15.7).

4 Management planning documentation may be most helpfully split into long, medium and short-term plans (see Box 15.8).

5 Medium and large FMUs almost certainly need to be divided into different zones, or working circles. It is unlikely that all objectives can be achieved at one time in all areas, so some sort of zoning will be necessary. Zoning can be used to identify areas with different primary objectives, such as timber production, tourism or recreation, NTFP collection or conservation. Some areas within the FMU may have legal protection or be inaccessible (such as watersheds or wetlands). Appropriate zoning should be based on information from environmental and social impact assessment, inventory, consultation and collaboration with stakeholders, HCVF surveys and conservation planning.

6 Much of the management planning information can be presented usefully in the form of maps. This is often a clearer and more useful form of presentation than describing prescriptions in text. Increasingly, many larger forest organizations are using geographical information systems (GIS) as a basis of management planning. However, this is by no means essential and paper-based maps may be equally good, particularly in smaller forest organizations and where computerized systems are not available.

WHAT DOES A MANAGEMENT PLAN CONTAIN?

The precise contents of the management plan documentation will vary according to local legal requirements, local forest management standards and guidance. However, management plans usually include the following elements:

Box 15.8 **Scales of management planning**

Many forest organizations find it useful to have two or three levels of planning over different timescales, which together make up management planning documentation. These might include:

- **A long-term plan** (sometimes called a strategic plan), which covers the entire FMU over the long term, such as an entire rotation, felling cycle or 20-year period. This summarizes the main plans and activities over the long term and will usually include:

  - long-term objectives of management;

  - information on the forest resources available and constraints to management, collected through forest inventory (see Section 15.4) and discussions with stakeholders (see Section 15.2);

  - zoning of the forest into areas (zones or working circles) with different primary objectives, such as timber production forest or protection forest;

  - a description of the silvicultural systems (see Section 16.3) to be used, calculation of the harvest rates and the division of harvestable area into annual operating areas (coupes) or compartments (see Section 15.5, Calculating Sustained Yield);

  - the design of the main transportation system (Forest Roading, Section 16.1);

  - the type of operations which will be undertaken, with an overview of when and where they will occur.

- **A medium-term plan** (sometimes called a tactical plan), which sets out the activities planned for the next five-year period in greater detail. The medium-term plan needs to be compatible with the long-term plan and budgeting. Objectives and targets to be achieved within the five-year period can be set out here, together with responsibilities for achieving them. If meeting an international standard is an objective of the forest organization, this might be planned and achieved over the five-year period covered by the medium-term plan.

- **An annual plan** (sometimes called the annual plan of operations), which details the precise activities to be undertaken during the year. This will specify what work needs to be carried out: for example, where harvesting will occur, which stretches of road will be built or maintained and when. The annual plan needs to be closely linked to annual budgets and to field planning for each operation. Annual plans are usually short and precise: rather than a long description of activities, they may be based on maps and compartment/coupe lists.

- basic information;
- management objectives;
- description of the local situation;
- future management.

Forest Stewardship Council-based standards also require a summary of the management plan to be made available to the public and this may also serve as a general summary of the plan. The public summary should cover all the major sections of the management plan, but exclude any confidential information. It should be written in straightforward language. Specific requirements for the contents of a public summary are included in the FSC Principles and Criteria (see Appendix 1.1) and/or the appropriate FSC-accredited national standard.

Basic information

This includes details such as the title page, date and timespan of the plan, location and size of the forest, authors and forest organization responsible for implementation. It may also include a space for formal approval of the plan, either from senior management of the forest organization, or by the state forest department or equivalent.

Management objectives

Setting feasible management objectives is partially dependent on the forest resources available and this section may be more easily written after the description of the local situation and resources available.

When deciding management objectives, different options need to be weighed up. For example, in natural forest these options might be the balance to be achieved between management for conservation and production, or timber versus NTFP harvesting. If good information about the possibilities of integration of different uses and objectives is not available, the precautionary principle should be applied: this will mean planning for a range of forest products or services and then closely monitoring the impacts of management. If the objectives seem to be conflicting, rational decisions must be made between them.

Defining objectives in the management plan can be integrated with the process of defining a sustainable forest management policy and setting related objectives and targets, described as part of the Environmental Management System (see Part Three). Box 15.7 shows an example of forest management plan objectives.

Description of the local situation

In order to plan the future management of the forest, it is essential to understand the existing local situation and resources available for management. The description of the local situation covers current conditions (and sometimes past management) of the FMU as well as the surrounding area. The aim is to provide the physical, socio-economic and environmental context in which the forest manager has to work, and sets the scene for the proposed future management options.

While the description of the local situation may need to refer to other information (such as inventory or growth and yield data) this can be included as appendices, referenced or summarized.

Future management

The future management part of the management plan sets out the proposed methods for achieving the objectives. This should draw together the information in the previous two sections (objectives and the local situation) to form the basis of prescriptions for future management of the FMU. This allows the forest manager to describe how the available resources will be managed to achieve the desired objectives. Different prescriptions may be needed for different zones of the forest, enabling the overall management objectives to be met.

Future management prescriptions may be divided into the long-term, medium-term and annual plans outlined in Box 15.8.

REVISING THE MANAGEMENT PLAN

Once prepared, the management plan needs to be revised periodically in order to take account of changing legislation and markets, new technology and the impacts of past management identified through monitoring. The long-term plan may need to be adapted to fit changing circumstances; medium-term and annual plans need to be rewritten regularly in any case. Some legislation and forest management standards require the management plan to be revised on a defined schedule, such as every five years.

To ensure that the revision of the plan is effective, there needs to be a mechanism for channelling information from both internal and external sources into the revision process. This means that staff

Box 15.9	**Addressing common problems with management plans**

The management plan exists but is not implemented. Plans may be inappropriate because they:

- spell out jobs and responsibilities for which there is little capability;

- require financial and human resources which are not available;

- present information in an inaccessible and impractical form;

- miss information from less obvious sources, such as needs and knowledge of forest workers or local people;

- result in limited understanding by forest workers or local people of what the plan is about;

- fail to generate energy, enthusiasm or commitment from these people – upon which implementation depends.

Consider trying the following:

1   Make sure that members of the forest organization at all levels are involved in writing the plan, especially when the plan is written by external specialists.

2   Set achievable and relevant objectives and targets.

3   Accept that some of the management plan is background information – it's there for reference and should be referred to as necessary.

4   Separate out the management plan into different volumes: perhaps keep practical information separate from background information; if information is available elsewhere, cross-reference to it rather than repeating it.

5   Make sure the practical information forms a working document: for example, leave blank spaces that need filling in and space for notes.

need to be aware when the plan is being reviewed and to have the opportunity to provide inputs; the revision process may also provide an opportunity to discuss management with stakeholders and seek external views on future forest management.

COMMUNICATING AND IMPLEMENTING THE PLAN

The management plan may need to be communicated to staff within the forest organization and to external stakeholders. To facilitate this, it may be useful to develop summarized versions of the important elements, together with clear maps, which can be understood by non-specialists. Documents may need to be translated into local languages.

The implementation of the plan is crucial. The objectives and prescriptions contained in the plan may need to be set out in procedures and guidelines which can be understood by field staff (see Chapter 12), introduced and emphasized through adequate training (see Section 16.5). This is particularly important in situations where staff turnover is high and continuity of management may be lacking.

Scale considerations! ♀

Management planning documentation is needed for all types and sizes of forest, but should be appropriate to the size and context of the enterprise and must meet requirements specified in the law and any relevant forest management standard. Where such requirements exist, they are sometimes different for small forests.

In many cases, if this is consistent with legal requirements, simple, annotated maps can provide much of the documentation needed for small forests where few operations are taking place.

15.4 Resource Surveys

WHY IS A RESOURCE SURVEY IMPORTANT?

A resource survey forms the basis for sustainable forest management. It is an assessment of the quantity and quality of forest resources available including timber inventory and growth and yield dynamics. A resource survey is important in both natural and plantation forests, but this section focuses on natural forests where survey work is often more complex. However, many of the principles are also applicable to plantations.

A forest resource survey in a natural forest may assess:

- timber resources, for example species, sizes, volume, number of trees;
- NTFPs, such as lianas, fruits, nuts, bamboo and wildlife;
- other aspects, such as soils, water sources and wildlife.

There are two basic methods of survey: static and dynamic. An appropriate survey is essential to the planning and management of any forest. In order to plan the sustained yield for the forest (see Section 15.5) the forest manager needs to know:

- *What resources currently exist?* For example, volumes of timber, dimensions and species. This information normally comes from static inventory – this consists of pre-harvest inventories, stock surveys and post-harvest inventories and is used to provide a 'snapshot' of the forest resources at a particular point in time.
- *How fast do the resources grow and what changes occur in response to management?* For example, rates of growth, mortality and changes in species composition. This information is collected from dynamic inventory. This is usually done through measuring permanent sample plots (see Section 17.2). This type of information is used mainly for long-term planning, calculating sustained yields and monitoring.

An example of these two types of mensuration and their use for SFM in Costa Rica is shown in Box 15.10.

At the scale of the whole forest management unit, the manager needs to know what resources are available (what species, what sizes, what quantities and where) in order to plan utilization, marketing and conservation. In the long term, inventory data about the existing forest resources, combined with data on growth rates, will provide an estimate of the time it will take for products harvested to regenerate. As discussed in Section 15.5, a prerequisite of sustainable forest management is that the removal of forest products does not exceed the rate of replacement. For this to be ensured, the existing resource, the harvest rate and the rate at which the resource is replaced after harvesting (growth rate) need to be known.

At a smaller scale, the forest manager may need more detailed information about the precise location and composition of the forest, to permit detailed planning of a harvesting unit. This would allow the forest manager to calculate the resource available and the costs of extracting it. It may also identify areas to be excluded from harvesting due to difficult access, low stocking levels, or for environmental or cultural protection.

Box 15.10 **The use of forest mensuration in SFM in Costa Rica[1]**

CODEFORSA is a private forest management organization that prepares management plans for owners of small blocks of natural forest (less than 500 ha) in northern Costa Rica. CODEFORSA uses three levels of forest inventory to determine the sustainable forest management planning system.

1 *Dynamic inventory:* A series of 36 permanent sample plots spread throughout the forests are measured once every two years in order to gather information about forest dynamics (growth, recruitment, mortality). This information has been used to construct a computer growth model (SIRENA) which is used to test the long-term sustainability of prescribed management regimes, forest by forest.

2 *A static sample:* As part of the planning process, a pre-harvest static sample of the forest woody biomass is taken (3 per cent for trees above 30 cm diameter and 0.75 per cent for those between 10 cm and 30 cm). This information is collected from a series of 0.3 ha rectangular plots, randomly distributed along forest survey lines (parallel access lines 75 m apart that extend over the whole of the area to be managed). The objective of the static sample is to help plan the *long-term management of the forest operation.* The static sample provides a 'snapshot' of the forest composition and structure. By feeding the data into the SIRENA growth model, management options can be tested for sustainability by simulating the forest's growth over a number of harvests. The static sample thus provides the basis for determining the felling cycle and the allowable cut and gives an initial indication as to whether post-harvest silvicultural treatments will be necessary.

3 *Stock survey:* In addition to the static sample, a 100 per cent enumeration of all the harvestable stems of commercial species above the minimum felling diameter of 60 cm is also carried out. The trees are measured, enumerated and their positions recorded as the survey team pass along the forest survey lines. At the same time a topographic survey is carried out which allows the forester to prepare maps of the forest operation showing contours, watercourses and the positions of the harvestable trees. The objective of this operation is *short-term harvest planning*, for example which of the harvestable stems to remove or where to locate the extraction routes. At least 40 per cent of the harvestable stems enumerated during the survey are not taken during the first harvest but are left as a seed source and to provide the basis for the second harvest at the end of the felling cycle (normally between 17 and 25 years).

WHAT IS REQUIRED?

There are many different ways of carrying out a forest resource survey, and almost as many textbooks describing those ways. Anyone considering doing a forest survey should consult one of the books listed in the Further Reading, Appendix 4.2, and preferably seek expert advice on the design. However, there are a few useful principles:

- Carrying out a forest inventory generally means measuring trees (although it might include NTFPs or other aspects, depending on the information required). A sample of trees is measured and the results extrapolated to the rest of the forest, taking account of the different forest types and their extent.
- Sampling a large enough proportion of trees to represent accurately the whole forest is costly. Depending on the objectives of the inventory, it may be possible to compile equally useful

1 S Maginnis, *Proyecto de Manejo Integrado del Bosque Natural*, CODEFORSA/ DfID, Costa Rica, 1997

information from existing sources: for instance, existing satellite imagery or aerial photographs, other partial inventories, comparison with similar stands, which have already been inventoried, and research results may be useful.

1 Types of inventory

The main types of inventory are the pre-harvest and post-harvest inventory (which can be static or dynamic) and the stock survey (see Box 15.10).

(a) Pre-harvest inventory

The purpose of the pre-harvest inventory is to collect information about the species present in the forest, how abundant they are, how they are distributed in the area and the range of sizes (size/class/distribution).

(b) Stock survey (100 per cent enumeration)

A stock survey is a 100 per cent sample of all harvestable trees over a certain diameter in size. It is carried out prior to harvest to assist with planning of harvest operations. Stock survey is normally done only for the part of the forest due for harvesting operations within the next one to two years (this may be the whole of a small forest). All commercial trees of harvestable size are identified, mapped and sometimes measured and tagged (see Box 15.11). The information collected can be used to produce a stock map which then forms the basis for planning which trees to fell, the location of extraction routes and felling directions. This is described in Box 16.5.

(c) Post-harvest inventory

This type of inventory is done by measuring trees in sample plots distributed throughout the forest area. Usually the plots cover somewhere between 0.5–2 per cent of the total area. Post-harvest inventory collects similar information to a pre-harvest inventory as well as information about any damage or changes caused by harvesting.

It is very important that this type of inventory work is done correctly to avoid bias and to ensure that the results give an accurate picture of the whole forest. An example of bias in post-harvest inventories is that crews tend to use extraction routes for access and set up sample plots near these routes where disturbance is greatest. This gives an inaccurate picture of the whole forest area.

2 Defining inventory objectives

The two most important questions to ask when defining inventory objectives are:

1 What do we want to know about the forest?
2 Why do we need to know?

The answers to these questions should be clearly related to the management plan objectives. Often the temptation is to collect as much information as possible because it might be useful, but collecting and analysing extra information costs time and money. Information is like any other production input – you need enough, but not too much or it reduces your margins.

3 Planning the inventory

To help plan the inventory, the following guidelines could be used:

- Check that the sampling method chosen is cost-effective. Inventories can be expensive, so it is important to find the system which provides adequately accurate information required at the lowest cost.
- Plan statistical analysis in advance so that the right information is collected in the right format.
- Ensure field teams are appropriately and effectively trained.

Box 15.11 Stock survey (100 per cent enumeration)

Most tropical natural forests are characterized by having a wide variety of species, of a range of sizes, irregularly located around the forest. A 100 per cent enumeration provides information about the species, sizes, numbers and locations of commercial trees in the harvesting area. Normally only commercial species of harvestable sizes are measured and recorded. It can also be used to record information such as steep slopes, swamps, rocky outcrops and inaccessible areas, to assist with planning of harvest operations.

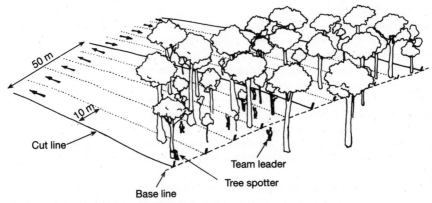

Basic steps (adapted from the CELOS system in Surinam) are:

- An enumeration team is put together, comprising a team of 'tree-spotters' and a team leader.
- Parallel lines are cut by forest workers through the forest at regular intervals along a baseline (for example at 50m or 100m intervals).
- Starting from the baseline, the team walks through each strip between two cut lines. The 'tree-spotters' walk abreast at regular intervals, checking the strip of forest on each side of them.
- Each tree of a commercial species and harvestable size located by the tree-spotters is identified, measured and sometimes tagged.
- The team leader, walking behind the tree-spotters, records the location, species, diameter at breast height (dbh), and sometimes height, damage, or other information, on a specially designed form.
- After the field work, the forms are compiled to produce a stock map for pre-harvest planning (see Box 16.5, Section 16.2) and data analysed to determine precise volumes, species and locations of trees to be harvested.

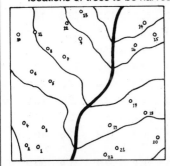

Tree number	Species	dbh
1	A	
2	B	
3	A	
4	C	
5	A	
6	A	
etc		

Sample stock map and accompanying information which may be used for pre-harvest planning

- Provide sufficient resources for logistics and to ensure the welfare of field teams.
- Make provisions for emergencies. This may mean having a vehicle available, radio communication and first aid equipment.
- Establish clear procedures for data collection. For example, ensure that teams are clear about plot demarcation, correcting for slope, measurement procedures, what to sample and what to ignore, and any additional information which needs to be collected.
- Use quality control procedures, such as independent checking (audit) of measurements and of species identification.
- Ensure there are enough personnel, time and funds, to manage and analyse the data.

4 It's not just the trees

Although trying to measure everything adds unnecessary costs, it must also be remembered that the forest manager may be interested in more than just trees. For example, sometimes stakeholder consultations or legal obligations may establish key non-timber values which need to be included in an inventory. Where High Conservation Values have been identified in the FMU, the survey may provide information on distribution, frequency and abundance.

Information on these other aspects of the resource may usefully be gathered during the survey. Setting clear objectives for the inventory is important in maintaining this balance. For example, it may be essential that any existing environmental damage is recorded, such as soil erosion, streambed erosion or oil spills. Other factors will depend on the site management objectives, for example, in natural forests, NTFPs, soils, wildlife or evidence of other users of the forest may all need to be recorded.

An inventory or detailed site survey prior to establishment of a plantation might be more comprehensive and could be carried out as part of the Environmental and Social Impact Assessment (see Section 15.1).

TABLE 15.1 Example of summary stock and stand table from Philippine–German Integrated Rainforest Management Project, Management Plan for 1993–2002

Number of stems and bole volume per hectare by commercial groups and diameter classes

| | Commercial Group | | | | | | | | | | | |
| | Common hardwoods | | Construction & furniture woods | | Light hardwoods | | Softwoods | | Others | | Total | |
Diameter Class	no/ha	m³/ha	no/ha	m³/ha	no/ha	m³/ha	no/ha	m³/ha	no/ha	m³/ha	no/ha	m³/ha
Regen-eration	912.7	–	853.1	–	64.0	–	2.2	–	4494.1	–	6326.1	–
10cm	128.0	4.2	130.9	1.9	16.0	0.2	0.0	0.0	592.4	8.7	867.3	15.0
20cm	20.8	4.3	26.4	3.1	5.7	0.7	0.0	0.0	86.9	10.3	139.8	18.4
30cm	8.3	4.6	8.4	3.0	3.4	1.2	0.0	0.0	28.5	10.3	48.6	19.1
40cm	4.5	5.5	3.4	3.2	1.1	1.3	0.0	0.0	7.7	7.0	16.7	17.0
50cm	4.9	10.0	1.8	3.1	0.4	0.9	0.0	0.0	3.4	5.3	10.5	19.3
60cm	4.2	12.6	1.0	2.6	0.1	0.4	0.0	0.0	1.9	4.4	7.2	20.0
70cm	3.4	14.6	0.7	2.6	0.0	0.1	0.0	0.0	1.0	3.3	5.1	20.6
80cm	1.9	10.2	0.4	2.4	0.0	0.0	0.0	0.0	0.4	1.7	2.7	14.3
90cm	1.0	6.6	0.3	2.8	0.0	0.1	0.0	0.0	0.2	1.2	1.5	10.7
100cm & up	1.1	14.8	0.5	5.9	0.0	0.3	0.0	0.0	0.2	2.2	1.8	23.2
Total	1090.8	87.4	1026.9	30.6	90.7	5.2	2.2	0.0	5216.7	54.4	7427.3	177.6

5 Data management and analysis

Forest inventory has no value if the data are not analysed and then fed into the management system. How data will be analysed, and by whom, needs to be considered at the planning stage; costs will need to be budgeted for before the inventory starts.

Recent years have seen the development of a range of software which can assist in analysing inventory results. For example, one software programme, TREMA (Tree Management and Mapping software for SFM), is designed for forest managers to improve tree management and mapping.

6 Final results of survey

The final result of a survey will depend on the initial objectives, but they should be fed into the forest management plan, providing information for calculation of the annual allowable cut (AAC). A standard timber inventory (for example, see Table 15.1) will typically result in a description of the timber resource of a particular area, broken down by forest types, species, sizes, total number of stems, number of trees per hectare, and volume per hectare. An estimate of the statistical reliability of the figures (which will depend on the original design of the inventory) is normally attached to each of these parameters.

15.5 Calculating Sustained Yield

WHY IS YIELD REGULATION IMPORTANT?

A prerequisite of sustainable forest management is that the removal of forest products does not exceed the rate of replacement (see Section 5.2). Without this basic balance, provided by yield regulation, sustainable forest management is impossible and the forest resource will gradually be depleted and degraded.

It is also of great practical importance to the forest manager, who needs to have predictable production levels each year. Where the forest is the major supply source for a processing facility, the customer wants to be assured that the supply will be fairly constant each year so that the full capacity of the processing facility can be supplied. In the absence of an infinite resource, the forest manager must therefore divide up the available forest into annual quotas.

WHAT IS REQUIRED?

Yield regulation is the practice of calculating and controlling the quantities of forest products removed from the forest each year to ensure that the rate of removal does not exceed the rate of replacement. If the main product being harvested is timber, yield regulation is often defined by the annual allowable cut (AAC). The AAC can be volume-based or area-based (see Figure 15.5):

- an area of land (hectares/acres) to be harvested annually – the area is fixed but the volume may fluctuate (though this would not be suitable where constant annual production levels are required); or
- a volume of timber (cubic metres) to be harvested annually – the volume is fixed but the the area of land from which it comes may vary from year to year, depending on the amount of standing timber and the distribution of timber species.

In simple terms the AAC is the volume (or area) of timber which is available divided by the number of years required until the next harvest. The production area is divided into annual coupes for harvesting: in theory, by the time the last block has been cut, the first block should be ready for harvesting again. The number of years into which the area can be divided (and hence the annual volume or area for harvesting) depends on the regrowth rate and the harvest intensity.

Box 15.12 **Addressing commons problems with inventory**

Inventory information is unreliable. Inventory crews are demoralized and are not providing accurate and timely results.
   Consider trying to:

1   Ensure that inventory crews' conditions in the field are adequate: in particular, ensure they have transport (if there is road/river access) and a member of the crew is trained in first aid.

2   Arrange field work to ensure that inventory teams have sufficient time between field trips to recover and that each field trip is of reasonable length.

3   Visit inventory crews sufficiently frequently to boost morale and carry out spot checks on quality of work.

4   Ensure the inventory crew have enough food and do not have to resort to hunting or return to base camp in order to supplement rations.

5   Consider financial incentives for providing accurate or timely results.

Although the principles are similar, the calculation of the AAC is slightly different for natural forest systems and plantations and they are described separately here.

1 Natural Forest

Calculations of annual allowable cut require as much information as possible about:

- The quantity and quality of the existing resource – including species composition, volume and their distribution per diameter class. Ideally this information is obtained through inventories (see Section 15.4).
- The rate of growth and mortality by species after harvesting at a particular intensity.
- The total production area (see Box 15.14).
- The levels of harvest or extraction which have been carried out already.
- Natural regeneration and ecological effects of harvesting. This information can be obtained through long-term monitoring and research.

Area-Based Yield Regulation

Requires even distribution of volume,
but less inventory data

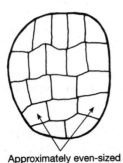

Approximately even-sized
annual felling coupes

Volume-Based Yield Regulation

Requires good inventory data but
works for uneven distribution of volume

Low density of timber: High density of timber:
large annual felling coupe   small annual felling coupe

FIGURE 15.5 **Volume and area-based yield regulation**

Box 15.13 **Addressing commons problems with annual allowable cuts**

Data on growth rates of commercially harvested species on which to base reliable estimates of the AAC are often lacking. In this situation, calculations have to be made based on data from similar forest types. A frequently quoted rule of thumb for growth rates in tropical moist forests is approximately 1 m^3/ha/yr. In many areas, more accurate information is available for a particular forest type and species.

Some tools exist to help with yield regulation in natural tropical forests. MYRLIN (Methods of Yield Regulation with Limited Information) is available free via the internet (see Appendix 4.1 and references in Appendix 4.2). It comprises three Microsoft Excel workbooks to assist in calculating the AAC when only limited research data are available on forest growth dynamics locally. The three tools comprise:

1   a stand table compilation module, to organize inventory data into stand tables;

2   a growth estimation module, to estimate increments and mortality rates for species;

3   a harvesting model that uses inventory and growth tables to estimate sustainable yield.

Where good data are not initially available, the problem must be addressed over the long term by:

- Initially *setting a conservative AAC*: underestimate the probable growth rate of the forest, and hence possible harvest levels, and overestimate the required felling cycle to ensure that future harvests will be feasible.

- *Monitoring growth rates* by establishing, measuring and analysing data from growth and yield plots (see Section 17.2).

- *Keeping records* of volumes and areas harvested so that future production can be estimated: even when reliable growth data become available, it cannot be applied if the original harvest levels from each area are not recorded.

- *Revising the AAC* – either up or down – as results of growth rate monitoring plots become available.

- *Carrying out sampling* (inventory) to determine when a compartment may be ready for a second harvest and to check that predictions of growth and yield are correct.

Some points to bear in mind when calculating the AAC:

- *Growth rates*: Growth rates are dramatically altered by logging. Even if the growth rates of the forest type in question are estimated from similar areas, the actual growth rates following logging must be measured. The rate of growth of the forest is determined from long-term studies using growth and yield plots (see Section 17.2).
- *True harvest intensity*: It is important to take account not only of the level of harvest itself, but also the level of damage caused to the forest during harvesting. For example, if the harvest level is set at 15 m^3 per hectare, but another 6 m^3 per hectare of timber is destroyed through trees being pushed over or damaged during felling and extraction, the total volume removed needs to be acknowledged as (and calculations based on) 21 m^3 per hectare.
- *Logging techniques*: The levels of damage from harvesting and subsequent post-harvest growth rates can be significantly affected by logging techniques. Careful logging (for example by directional felling, use of winches and advanced planning for all skid trails) minimizes damage and ecological disruption to the forest and potentially shortens cutting cycles. Studies in Cameroon, Costa Rica, Guyana, Indonesia and Surinam have shown that damage to the forest can be substantially reduced by limiting the length of skid trails and increasing the use of the winch (see Box 16.8).

Box 15.14	Defining productive forest area

An error which is sometimes made in calculating annual allowable cuts is to forget to exclude all non-productive forest from the calculation. In addition to naturally non-productive forest such as steep areas and wetlands, and protected forest, such as conservation areas and buffer zones along rivers, deductions in area also need to be made for permanent roads, semi-permanent extraction routes, log yards and other infrastructure. This can be significant. For example, estimates suggest that with ground skidding systems, up to 30 per cent of the forest area may be occupied by roads, log yards or skid trails.[2]

AN EXAMPLE:

The forest manager has to estimate the productive area of an organization's 10,000 ha of mixed natural forest. Inventory data tell the forest manager that 10 per cent of the forest is unproductive, or inaccessible. 10 per cent has been set aside as conservation areas. In addition, a further 5 per cent consists of rivers, swamps and streamside buffer zones. Another 10 per cent of the forest is covered by roads, skid trails and log yards, calculated as an average per hectare.

Utilizable forest is the area that can be harvested. This excludes inaccessible and unproductive forest, log yards, roads, rivers, buffer zones and all conservation areas.

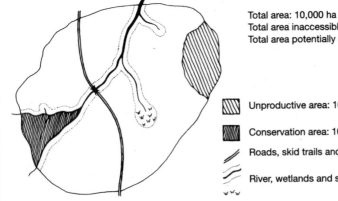

Total area: 10,000 ha
Total area inaccessible and unutilizable: 3500 ha
Total area potentially utilizable forest: 6500 ha

Unproductive area: 1000 ha

Conservation area: 1000 ha

Roads, skid trails and log yards: 1000 ha

River, wetlands and streamside buffer zones: 500 ha

- *Operational control*: In order to make sure that harvest levels on the ground are in line with those predicted, an effective means of operational control of harvests needs to be in place. The forest manager needs to ensure that logging crews do not concentrate on high-value or easily accessible trees. Adequate numbers of seed trees must also be left.
- *Silviculture*: In selection systems, post-harvest silvicultural treatments can have significant impacts on growth rates and thereby reduce the length of the cutting cycle (see Section 16.3).
- *Ecological relationships*: Perhaps the least known factor affecting yield regulation is the ecological inter-relationships of the forest and the effect they may have on sustainable forest management. Particularly in tropical forests, little is known about the details of forest ecology. Interdependencies among animals and plants, and plants to plants are common, and sometimes very specific. For example, some species of orchid have their 'own' species of bee to pollinate them. Removing a keystone species from the ecosystem might have knock-on effects for other species (see Box 6.1).

2 LA Bruijnzeel and WRS Critchley, *Environmental Impacts of Logging Moist Tropical Forests*, UNESCO, Paris, 1994

Box 15.15 **Addressing common problems with changing markets and sustained yields**

Systematic working of the production area, using a calculated AAC, requires that annual blocks are harvested and then closed to further felling. Blocks should remain closed until adequate regeneration and regrowth has occurred to support a further harvest: if the AAC has been correctly calculated and implemented, this will be at the end of the felling cycle. Changes in markets often provide the temptation to re-enter harvested blocks before the next scheduled felling: species which at one point in time were regarded as non-commercial and were therefore not harvested may become commercially valuable. Roads and infrastructure already exist to support further harvesting.

There is no simple solution to this problem: re-entering a compartment should not be allowed until it is silviculturally advisable and in accordance with SFM practices to do so on the basis of adequate inventory and growth and yield data.[3]

All the above factors need to be combined to arrive at a balanced and feasible harvest level. For example, inventories may show that 100 m³ per hectare of desirable timber species, of harvestable size, exist in an area of forest, but ecological research may also show that removal of all 100 m³ on each hectare would severely disrupt the forest ecosystem, affecting its ability to regenerate and remain a productive resource. In the interest of sustainability, the harvest level must be set at an ecologically safe level.

2 Plantations

For plantations the existing resource and growth rates are often better known and documented, and the ecological inter-relationships within the plantation are simpler than in natural forest. In simple terms, this allows the AAC to be calculated as the annual increment over the whole forest operation. For example, if the increment is 35 m³/ha/year over a plantation of 100 ha, the total increment of the plantation is 3500 m³/year. In theory it should be possible to harvest 3500 m³ of timber from the plantation each year on a sustainable basis. The size of area from which this comes, the annual felling coupe, depends on the amount of timber standing on each hectare. The rotation length will depend on how fast the trees grow, and at what point they reach valuable, utilizable sizes. In practice, therefore, felling may take place less frequently than once a year as insufficient numbers of trees will have reached harvestable size each year.

Alternatively, if a plantation is attached to further processing plant, such as a pulp mill, it may be important to calculate the annual demand for products from the plantation. With known yields per hectare, the area needed to supply that demand can be calculated.

Example of calculation of plantation areas required to fulfil demand:

- A pulp mill requires 500,000 m³ of wood annually;
- The plantation, at maturity, yields 300 m³ per hectare;
- 1667 ha of plantation must be harvested annually to supply the mill.
- If the species in question requires 15 years to come to maturity then the plantation must be 1667 ha x 15 years = 25,000 ha in total area.

Further references on calculating sustained yield can be found Appendix 4.2.

3 H ter Steege et al, *Ecology and Logging in a Tropical Rain Forest in Guyana*, 1996. Tropenbos Series 14, Stichting Tropenbos, ISBN 90-5113-026-0

15.6 Developing a Conservation Strategy and the Management of High Conservation Value Forests

WHY IS A CONSERVATION STRATEGY IMPORTANT?

All forests embody a number of environmental and social values, such as acting as a habitat for many plant and animal species, protecting soils against erosion, or including areas of cultural significance. This is why ensuring the conservation of plants and animals, and the maintenance of other environmental and social values are important parts of routine forest management. When the environmental and social values of a forest are considered to be of outstanding or critical importance, they can be called High Conservation Values (HCVs) and the forest may be designated as a High Conservation Value Forest (HCVF).

Many of the issues discussed in this handbook relate to conservation, but this section looks closely at the additional factors that may be required in developing a strategic approach to conservation and the management of High Conservation Value Forests.

WHAT IS REQUIRED?

A conservation strategy or plan provides a systematic basis for the management of the environmental and social values of the forest. The conservation plan needs to be compatible with, and incorporated into, the management plan. Actions specified in the conservation plan may affect the way routine forest operations are carried out, as well as defining conservation areas to be excluded altogether from forest operations. This will have an impact both on the implementation of operational plans and procedures, and on long-term management planning.

Formulating a conservation plan may require specialist skills which are not available within the forest organization. This is particularly important for large organizations managing substantial forest areas, whether natural or plantation. Support and practical help can be obtained from a variety of sources including:

- forest departments;
- wildlife departments;
- local universities;
- specialist consultants;
- environmental (and social) NGOs.

However, before contacting such organizations, it is useful for the forest manager to have some idea about what is needed and this is outlined below. The most important elements of a conservation strategy are:

- protecting conservation areas;
- protecting rare, threatened and endangered species;
- identifying and controlling risks to biodiversity (for example, control of hunting, fishing and collecting, fire control, prevention of land invasion);
- managing High Conservation Value Forests.

Protecting conservation areas

Forest management will always have some impact on the forest as a natural ecosystem, however good it is. Therefore, some areas need to be set aside which are completely protected, even where High Conservation Values are not present. This will ensure that even the species that are the most sensitive to disturbance have a chance of surviving – as well as to provide a baseline against which the impacts of forest management can be assessed.

Box 15.16	**Virgin jungle reserves in Peninsular Malaysia[4]**

Virgin jungle reserves (VJRs) of Peninsular Malaysia are a network of small protected areas mostly located within selectively logged natural forest. The network was established in the 1940s and range from 2 to 2744 ha in size. The two main aims are to provide:

1   undisturbed areas of natural vegetation for non-destructive scientific research, in particular control sites for forest research;

2   protected examples of the country's national heritage.

Although the VJRs were originally conceived to act as gene pools or sources of genetic replenishment for surrounding logged forest, they are probably insufficient in number and size to perform this function.

Recent research has shown that the VJRs in Peninsular Malaysia have a greater probability of survival and utility for biodiversity conservation if they are greater than 260 ha (the original lower limit size) and occupy more than one compartment. They should be compact in shape and located centrally in the Forest Reserve (FMU) to reduce disturbance.

The total amount of land that should be protected will vary depending on the forest type, the area and local legislation. For example, Indonesian forest concessions are required by law to set aside and protect blocks of forest as a genetic reserve. The Ministerial Act on the Management and Utilisation of Biodiversity Conservation Plots in Production Forests (1999), requires concessions to maintain an area of between 300 and 500 hectares, representing different ecosystems in the FMU.

Similarly, in the southern and centre-western areas of Brazil, plantations (and other crops) established on previously forested land must keep at least 20 per cent of the area covered by natural vegetation in legal reserves. Although sustainable forest management is permitted in these areas, they are usually maintained by plantation forestry companies as conservation areas. In the Amazon region, 80 per cent of the FMU must be maintained as a legal reserve, where forest management is permitted, but conversion is prohibited. Requirements for the size and management of legal reserves are set out in the Brazilian Forest Code (1965).

The location of conservation areas is crucial in determining how useful they are. One important factor, in allocating forest within the FMU as conservation areas, should be to aim for areas which are *representative* of the vegetation types found within the FMU. Most medium and large forest organizations have several different types of forest within the managed area that they should aim to protect. For example, this may include lowland forest, upland forest or riverine forest. If only the riverine forest is protected, it is not representative. It is tempting to include only unproductive forest within conservation areas (such as steep slopes or low productivity forest), but including areas of productive forest is essential in order to be representative.

Ensuring that conservation areas are representative or most usefully located demands a thorough understanding of the forest types within the area. If data are not available it may be necessary to bring in a local expert or external specialist to assist in identifying forest types and then locating areas which should be conserved.

Attention should also be paid to the location of conservation areas to ensure that, whenever possible, they are linked together, allowing animals to move between the areas. These links are known as wildlife corridors. Ideally they should be away from roads and farm crops. Where patches of High Conservation Value Forest have been identified, additional conservation areas may most usefully be used to provide wildlife corridors linking them together.

4 RK Laidlaw, 'Virgin jungle reserves of Peninsular Malaysia: Small protected areas in logged forest', *Commonwealth Forestry Review* 77(2), 1998.

Protecting rare, threatened and endangered species

Special measures need to be taken in forest management to ensure the protection of rare, threatened and endangered species. These may include both animals and plants; they may be rare or threatened at a local, national or international level. Rare, threatened and endangered species can be considered as species that are either:

- protected by national or local legislation; or
- listed by international conservation assessments as being threatened or endangered. For example, the IUCN produces global 'red lists' of threatened plant and animal species, and CITES lists species threatened by trade.[5]

Although most forest management standards concentrate on the protection of rare species, there may be other species within the forest that it is important to conserve (see Box 15.17).

From 2005, locally adapted FSC standards produced by FSC national initiatives, must include a list of, or make a reference to official lists of endangered species in the country or region where the standard is to be used.

Rare, threatened and endangered species in the FMU need to be identified and protected. This will probably overlap with the other elements of the conservation strategy. For example, rare species may be a target for commercial hunting, which should be controlled in the FMU. The protection of a particular species will involve not only protecting the species itself from hunting or collecting, but may also require the protection of its habitat by setting aside conservation areas.

Where there is a significant concentration of rare, threatened or endangered species in the FMU, this would be considered a High Conservation Value and the forest should be managed as a High Conservation Value Forest.

Controlling threats to biodiversity

An important part of a conservation strategy is to ensure that management does not threaten the survival of rare, threatened and endangered species or forest types. There are a number of common threats to biodiversity, which need to be controlled as part of the strategy. Not all these threats are applicable or potentially severe in any particular forest, but they need to be considered. Common threats include:

- *Inappropriate levels of hunting, fishing and plant collecting within the forest area.* Where there is no legal or customary right to hunt, fish or collect, forest organizations sometimes choose to ban such activities. However, unilaterally banning hunting may have severe negative social impacts on local communities where there is a dependence on hunting. Development of alternative options (such as employment as guides, in eco-tourism ventures, etc) may be possible. Providing people with substitutes for hunting is more likely to be effective than trying to enforce a ban. (See also Box 19.3 for a discussion of determining hunting rights and responsibilities.)
- *The incursion and spread of fire*, particularly in areas where fire is not part of the regular dynamics of the forest. This is often particularly a problem when the forest area is surrounded by other land uses where fire is part of the management regime (for example, many tropical pastures). Minimizing the risk of fire incursion may include a combination of measures, such as maintaining fire breaks, developing a fire vigilance and response strategy and coordinating and collaborating with neighbouring properties.
- *Illegal invasion and settlement* within the forest. Mitigating the risk of illegal land occupation is difficult, but can be helped through fostering good relations with local communities. This can both reduce the risk that the forest will be chosen for invasion and potentially provide warnings that such events are likely to occur. Once an invasion has happened, prompt action from the

---

5 See Appendices 3.2 and 4.1 for further information.

Box 15.17 **Some characteristics which can be used as a basis for choosing species of high conservation priority**

Characteristic of species	Higher conservation indicated by:
Global distribution	Globally rare
National distribution	Nationally rare
Ecological range	Narrow range of habitats
Ecological specifics	Species of undisturbed habitats
Human considerations	Threatened by use, or closely related to economically important species. 'Cute' or otherwise popular
Interactions with other species	'Keystone' species, on whose existence many other species depend
Taxonomic	Only species of its 'type' (eg only species in family or genus)
Environmental association	Associated with fragile soils or otherwise known to play an important role in ecosystem

---

authorities is likely to be required to resolve the issue and minimize damage. Keeping clear communication channels with the appropriate authorities is essential.

- *Forest management activities*, such as inappropriate harvesting or silvicultural treatments. The forest manager has a great degree of control regarding harvesting techniques and the application of silvicultural treatments, and can choose an option that is both financially rewarding and reduces threats to biodiversity.

High Conservation Value Forest (HCVF)

Almost all standards for sustainable forest management contain requirements for the protection of important wildlife habitats, biodiversity and ecosystem service functions (such as the protection of the watershed, soil conservation, or enhancement of natural biogeochemical cycles), as well as the preservation of important social elements for forest-dependent communities. The HCV concept is based on the idea that, when a forest contains a value of outstanding significance or critical importance, there need to be extra safeguards to ensure that the value is not being degraded or otherwise negatively affected by management.

The concept of High Conservation Value Forest was initially developed by the Forest Stewardship Council (FSC) for use in forest management certification, but the concept has now been incorporated more broadly in other conservation schemes. For example, the HCVF approach is increasingly being used by timber-purchasing companies as part of their responsible purchasing policies.

The key to the concept of the HCVF is the identification and management of High Conservation Values (HCVs). The area of forest needed to maintain these values is the High Conservation Value Forest (HCVF). A High Conservation Value Forest may be a small part of a larger forest, for example a riparian zone protecting a stream that is the sole supply of drinking water to a community, or a small patch of a rare ecosystem. In other cases, the High Conservation Value Forest may be the whole of a forest management unit, for example when the forest contains several threatened or endangered species that range throughout the forest. Any forest type – boreal, temperate or tropical, natural or plantation can potentially be a High Conservation Value Forest, because the HCVF designation relies solely on the presence of one or more High Conservation Values.

Box 15.18	**The six High Conservation Values**

HCV 1:   Globally, regionally or nationally significant concentrations of biodiversity values (eg endemism, endangered species, refugia). For example, the presence of several globally threatened bird species within a Kenyan montane forest.

HCV 2:   Globally, regionally or nationally significant large, landscape level forests, contained within, or containing the management unit, where viable populations of most if not all naturally occurring species exist in natural patterns of distribution and abundance. For example, a large tract of Central American lowland rainforest with healthy populations of jaguars, tapirs, harpy eagles and caimans as well as smaller species.

HCV 3:   Areas that are in or contain rare, threatened or endangered ecosystems. For example, patches of a regionally rare type of freshwater swamp forest in an Australian coastal district.

HCV 4:   Areas that provide basic services of nature in critical situations (eg watershed protection, erosion control). For example, forest on steep slopes with avalanche risk above a town in the European Alps.

HCV 5:   Areas fundamental to meeting the basic needs of local communities (eg subsistence, health). For example, key hunting or foraging areas for communities living at subsistence level in a Cambodian lowland forest mosaic.

HCV 6:   Areas critical to local communities' traditional cultural identity (areas of cultural, ecological, economic or religious significance identified in cooperation with such local communities). For example, sacred burial grounds within a forest management area in Canada.

There are six main types of HCV, which together encompass exceptional or critical ecological attributes, ecosystem services and social functions, shown in Box 15.18.

There may be some trade-offs between the allocation of representative conservation areas and the identification and management of High Conservation Value Forests. For example, if an FMU contains a large area of a threatened ecosystem (HCV 3) where forest management activities cannot be carried out, this may also fulfil the function of a conservation area. However, if the forest contains an HCV that can be maintained at the same time as forest management activities are carried out, separate, additional conservation areas may be needed where no management occurs.

Nevertheless, designating a forest (or part of a forest) as an HCVF does not automatically preclude management operations such as timber harvesting. It does mean that management must ensure that the HCVs are maintained or enhanced. Careful attention needs to be paid to planning and implementing any forest management activity that could potentially affect the High Conservation Values.[6]

DEVELOPING A CONSERVATION STRATEGY

The four elements described above need to be brought together as part of a conservation strategy. It is important that the conservation strategy is also adequately integrated within overall management planning and operational procedures.

---

6 For more detailed information on HCVF, see 'The High Conservation Value Forest toolkit, Part 3 – Identifying and managing High Conservation Value Forests: a guide for forest managers'. Available at www.proforest.net

The process of identifying, managing and monitoring conservation values can potentially be quite time-consuming. This is particularly the case if the forest is thought to contain HCVs, where differences in stakeholder interests and their interpretation of the conservation values may need to be resolved. Therefore, it is very important for forest managers to begin the process by trying to investigate the interests and needs of known and potential stakeholders. Forest managers also need to be familiar with national guidelines or standards relating to the conservation of globally significant biodiversity or environmentally sensitive areas in their forest.

Where appropriate, forest managers may need inputs from experienced ecologists or social advisers to provide technical advice on biodiversity evaluation and stakeholder consultation processes, as well as on developing a framework for management and monitoring.

The three main steps in developing the conservation strategy are:

1 assessment of conservation values in the FMU;
2 developing a management strategy;
3 monitoring and feedback.

Assessment of conservation values and threats in the FMU

The first step in developing a conservation strategy is to find out what the forest contains of conservation value, where it is and what it needs protection from. This means the forest needs to be assessed to identify:

- Rare, threatened or endangered species that should be protected.
- Forest types present in the FMU that may contribute to conservation areas.
- Threats to biodiversity (including hunting, fishing and collecting activities, danger of fires, potential for land invasions and inappropriate harvesting and silvicultural activities) that may need to be controlled.
- High Conservation Values that will need special attention in forest management.

Field surveys are not always the best way to assess conservation values

The first response of many forest managers to this is to conduct or commission field surveys. However, such surveys are not always the best recourse: they can be costly and result in little more than a list of which species are present, which is not always useful in terms of management planning. Assessment of conservation values can be approached in a five-step process:

Step 1: Assessment of the landscape context This only applies to medium and large forest areas, and involves developing an understanding of whether the forest area plays an important role in the wider landscape. Examples of forest areas that might be expected to play an important role in biodiversity conservation at a landscape level include those that:

- border with a protected area;
- connect two or more protected areas;
- comprise, or are part of, a large block of forest in a country or region where large areas of forest are rare.

Information on this can be gained through studying maps, satellite imagery, broad-scale conservation assessments and talking to conservation experts.

Step 2: Assessment of forest types This step involves finding out what forest (or other vegetation) types are present within the forest management unit. Much of this can be determined from information that the forest manager already has available, such as vegetation maps, data from permanent sample plots or timber inventory. Ground surveys or fly-overs may be needed to provide some additional information.

Once the vegetation types in the management unit are known, they need to be assessed to determine whether any of them are threatened or have particular importance for endangered species. This information may also be important for identifying representative conservation areas. This assessment can be made by consulting existing literature or conservation experts.

Step 3: Identification of species and habitats This step involves first gaining an idea of what rare, threatened or endangered species could potentially occur within the forest area and then confirming whether they or their habitats are actually present.

An initial check of local, national or international lists of threatened species will show which locally known species should be counted as rare, threatened or endangered. It is also important to consider whether there are any habitat features in the forest that are important to these species. These might include, for example, large, hollow trees that can provide nesting or roosting sites for threatened birds or bats, salt licks, small lakes or other water bodies.

Information on whether any of these species or their habitats actually occur in the forest can be gained from maps, consultation with people who are familiar with the forest, and field surveys. A great deal of information can be gained from talking to people who are familiar with the forest. For example, in South America, forest workers are often aware if the forest contains jaguar and tapir because they see signs of these animals as they go about their normal field activities. Additional information can be gained from the conservation literature and from local conservation experts.

In many cases, this will be sufficient and the forest manager will now be aware of which rare species and important habitat features are likely to be found in the FMU. In some cases, however, further information may be required. This might happen, for example, when the species in question are not easily observed or identified by non-specialists. In this case, it may be necessary to undergo detailed field surveys, focusing on the species groups in question. Local universities and conservation experts are often good sources of help for this kind of difficulty.

Step 4: Identification of threats It is a relatively straightforward task to list the potential threats to biodiversity that could affect the FMU. This should only include those threats over which the forest

Box 15.19 **Sources of information for assessing conservation values in the FMU**

A: Background checks and information

- Information can be obtained from staff and local communities about the conservation values and species that have been seen in the forest.

- Consult existing sources of information to see whether all or part of the FMU has been identified as being of outstanding value for any of the six types of HCV or equivalent local designations.

- Check existing surveys and information (eg in local universities, government or environmental groups) about rare, threatened and endangered species in the FMU or surrounding area.

- Check existing surveys or inventories on forest/vegetation types.

B: Field survey (as necessary to supplement background information)

- Assess forest/vegetation types, if information is not already available.

- Confirm or clarify presence of HCVs thought likely to be present from background checks.

- Identify rare, threatened and endangered species present in the FMU.

C: Stakeholder consultation (throughout assessment period)

- Identification of High Conservation Values 5 and 6 of importance to local communities (ie areas fundamental to meeting basic needs and areas critical to traditional cultural identity).

- Consultation to determine whether other conservation values identified are critical enough to be considered as *high* conservation values.

- Identification of, and initial discussions about managing, hunting, fishing and collecting and other potential threats to biodiversity.

---

management has some direct control: listing global climatic change or air pollution is not useful from a management perspective! It is also helpful to prioritize which of these threats are most important within the FMU, in terms of how likely they are to occur and the impact they could have if they do occur (for example, affecting a small group of species only, or damaging whole ecosystems). This sort of analysis may already have been carried out as part of the ESIA (see Section 15.1).

Step 5: Identifying HCVs and delineating HCVFs Where a conservation value is of critical or outstanding importance, extra precautions need to be taken to ensure that the values are not degraded or lost. Because of the relative newness of the HCVF concept, there is often confusion about how HCVs should be identified and an appropriate area of forest delineated. National FSC standards are increasingly incorporating a definition of what constitutes a High Conservation Value Forest within Principle 9 of the standard. However, this is not always clearly defined and many forest managers work in countries where there is no national standard. In such situations, the forest manager will need to take the initiative in deciding what should be considered a High Conservation Value.

The assessments already done (steps 1–3 above, and consultation with local stakeholders) will provide much of the information needed to decide whether any of the conservation values identified are exceptional. The following process may be useful:

Box 15.20 **Examples of HCVs identified from existing information**

For HCV 1, check whether the FMU contains or borders with any Protected Areas or has been identified by a conservation assessment as being of outstanding importance (eg BirdLife International's 'Important Bird Areas').

For HCV 4, check whether all or part of the FMU is identified by government (or other) classifications as being of high priority for watershed protection, erosion control, etc.

For HCV 5, talk with local people about whether and how communities use the forest for resources that are likely to be critical to them (eg food, medicine, water, etc). This may be combined with on-going consultation processes (see Section 15.2).

- Develop a simple checklist of High Conservation Values that may be present in the FMU, under each of the six broad headings provided in Box 15.18. (For example, slopes steeper than 35 per cent on clay soil, greater than 2 hectares in area might be considered critical for erosion control (HCV 4).)
- Consult existing sources of information to see whether all or part of the FMU has been identified as being of outstanding value for any of the six types of HCV or equivalent local designations. Much of this information may already have been identified through other management activities such as ESIA (see Section 15.1), as it is also required for other aspects of forest management.
- Some of these sources of information (such as conservation planning maps or watershed protection zonings) will allow a quick decision that part or all of the FMU definitely is a High Conservation Value Forest. Others will indicate that the FMU may contain one or more HCVs. More information will be needed about what the FMU contains: this would include field checks and consultation, for example to find out whether or not the FMU contains a number of rare species, or whether the FMU is the only appropriate source of food for nearby communities, etc.
- At this point, it will be necessary to consult with stakeholders, to determine whether or not the values found in the FMU are sufficiently critical to constitute an HCVF. For example, are there so many rare species, or are some of the species so rare, that the forest contains HCV 1? For HCVs 5 and 6, it will always be necessary to make this decision with the participation of local communities.
- If it is decided that the forest does contain an HCV, the final step is to delineate the area of forest required to maintain or enhance it (ie the High Conservation Value Forest). Sometimes this decision will be relatively easy. For example, where the FMU contains a small discrete patch of a rare ecosystem, or where a local community has a specific sacred area, these areas can be delineated as the HCVF. In other cases, further information or consultation with experts may be needed, for example, to find out the habitat requirements of a group of rare species.

Developing a management strategy

The assessment of conservation values and potential threats should give a clear picture of conservation priorities for the forest management unit. This assessment should be used to develop an FMU-specific management regime, which is integrated with the overall management of the forest, both in terms of landscape level planning or zoning and the planning and implementation of specific operations (see Box 15.21). Consultation with stakeholders is likely to be necessary, especially where HCVs are known or thought to exist.

The conservation strategy should be based around a combination of approaches, which are likely to include the following:

Box 15.21 **Using the assessment of conservation values**

The conservation strategy must take into account all of the important aspects identified during the assessment of conservation values. For example:

- *Landscape context* – a broad strip of forest may be designated for very low harvest intensity to allow animals to move between a neighbouring national park and a large area of intact forest that borders the forest management unit.

- *Assessment of forest types* – complete protection may be given to a block of forest that contains both a patch of rare limestone forest as well some of the more representative lowland forest that covers most of the forest management unit.

- *Identification of species and habitats* – riparian protection zones could be designed to provide connectivity between the two areas above and large, hollow trees could be identified in timber inventories and retained as nesting or roosting sites during harvesting.

- *Identification of threats* – hunting of an endangered species could be controlled through a combination of consultation and participation with local communities and regular patrols.

- **Protection** measures for specific sites, such as the definition of conservation areas and buffer zones. This might also include controlling activities which degrade protected areas, such as the hunting of rare species. However, it is important to take into account the importance that some hunting, fishing and collecting may have for local communities and consider the best ways to control inappropriate activities. Box 15.22 discusses some ways of controlling hunting on the FMU.

  The protection of rare, threatened and endangered species and their habitats is likely to be a core part of the conservation strategy, where they exist. Where possible, these protection measures should be based on good knowledge of the ecology of the species in question. However, even where only inadequate survey information and ecological data are available, some common-sense measures can be taken to protect rare species (see Box 15.23).

- **Modifications or constraints on forest operations.** The way in which forest operations are carried out may need to be changed in order to reduce negative impacts on the environmental and social values of the forest. These modified management actions need to be incorporated into normal forest activities and are not necessarily site-specific. They might include, for example, modifications to the way roads are built or maintained to protect watercourses from siltation, or the use of reduced impact harvesting and extraction techniques.

- **Restoration** may be needed where the forest requires some remedial action, such as the removal of alien species, enrichment planting in heavily logged areas or the restoration of degraded riparian areas.

- Where **High Conservation Values** are identified, specific management regimes will need to be defined in the conservation strategy, which ensure that each HCV is maintained or enhanced. The area of forest necessary to maintain or enhance the HCV needs to be determined (ie the High Conservation Value Forest, or HCVF) and specific management prescriptions put in place. The way in which forest operations will be modified should be clearly identified and documented.

  Where HCVs have been identified in the forest, a precautionary approach must be taken. This means that if there is any doubt about how management will affect the HCV, it should be assumed that the impact will be negative and protection should be the preferred option until more information is available.

Box 15.22 **Approaches to controlling hunting**

Control of hunting may be achieved by a combination of:

- *Collaboration* – Working together with local government officials such as environmental or wildlife officers. Where possible, involve local environmental groups or other interested parties. Work together with communities and staff to build a commitment to conservation.

- *Information* – Ensuring that all staff and local people know that hunting, fishing and collecting are banned. Put up signs and produce publicity to explain that such activities are not permitted within the forest area.

- *Education* – Ensuring that all staff and local communities understand why the activities are banned. Education activities may include running classes for staff or local people, producing literature and talking to the school children.

- *Enforcement* – Where possible, controlling access to the forest. Larger companies sometimes use forest guards to patrol the forest area. Train all staff to report evidence of any banned activities.

Where there are legal or customary rights to hunt, fish or collect plants, then it is important to control activities. Many of the techniques described above can be used to ensure that rare and endangered species are protected even when other species are hunted, and that the hunting or fishing activities do not threaten other species. For example, fishing with a net could be prohibited but fishing with a line permitted, since the latter has a much lower impact. Ideally it is best to work with the hunting/fishing and gathering groups to agree management rules, and to encourage a degree of self-regulation.

The overall conservation strategy for the FMU will probably include a combination of the approaches detailed above. It is important that the rationale for adopting particular conservation measures is clear and appropriate to the site. The conservation strategy needs to be integrated into the broader forest management system. Forest management planning documentation (see Section 15.3) should include details of the conservation strategy and operations to be undertaken. Likewise, operational procedures need to incorporate any conservation considerations to ensure that they are routinely put into practice.

Box 15.23 **Some common-sense protection measures for rare species**

- Mark and maintain buffer zones around the nests of rare birds or dens of animals.

- Ensure that some areas of the habitat of a rare animal or plant are totally protected. This may require some improvements to habitat, for example leaving dead or snagged trees.

- Stop operations close to a rare bird or animal during the breeding season.

- Ensure that rare trees or shrubs are not disturbed during harvesting.

- Educate staff and local communities not to hunt a rare animal or collect a rare plant.

Monitoring and feedback

As the conservation strategy is implemented, it is also important to monitor the effect. As with all monitoring, a mechanism is needed to feed back the results of monitoring into the revision of plans. The monitoring system is likely to include a combination of:

- operational monitoring to ensure that plans are being put into practice with the desired effect; and
- strategic monitoring to examine the longer-term impacts of management and to check that conservation values are being maintained or enhanced (see Chapter 17 for more information on monitoring).

The information that was collected during the assessment of conservation values of the forest may well serve as a baseline against which the impacts of management can be compared. Similarly, where conservation areas have been completely protected, they may provide a baseline for comparison with managed areas of forest.

Where the forest contains one or more HCV, a specific monitoring programme is needed. The results of HCVF monitoring need to be checked and fed back into management planning at least annually.

Scale considerations!

The implementation of HCVF requirements in small forests is specifically addressed in Appendix 2.2.

In most cases, small-scale organizations will not need to consider all of the five steps of conservation assessment outlined above and the conservation strategy will be more straightforward. Conservation will normally be limited to the protection of specific sites and rare species, together with some constraints on forest operations. Measures are likely to include the 'common-sense' protection measures outlined in Box 15.23.

If small-scale organizations work cooperatively, it may be possible to implement a conservation strategy covering an agreed percentage of the total forest area held by all of the organizations.

16 Implementation in the Forest

Implementation looks at the practical forest management activities in the forest and focuses on the operational level. It is guided by the overall planning described in Chapter 15. Implementation of forest management activities will be assisted and supported by the implementation elements of the EMS described in Chapter 12.

The key elements of implementation covered here are:

- forest roading;
- harvesting and extraction;
- silviculture;
- chemicals and pest management;
- training.

16.1 Forest Roading

WHY IS FOREST ROADING IMPORTANT?

Forest roads are an essential part of most forestry operations. They provide access to the forest for management and harvesting operations, and extraction routes for products from the forest. Roads may also be important for other stakeholders for forest-related livelihood or business purposes. Areas which were previously inaccessible to people are made accessible. The forest, and people who live in and around it, are affected by roads. As a result forest roading can have important implications for the sustainability of forest management (see Box 16.1). Use of good roading practices minimizes negative impacts. In general, forest roading includes the planning, design, construction, use and the maintenance of forest roads.

WHAT IS REQUIRED?

Forest roads are major engineering structures. Constructing them requires the removal of the forest cover and movement of large quantities of earth and surfacing materials. Drainage patterns may be changed or blocked and new areas of soil are exposed to erosion.

Guidelines should be developed and implemented to ensure that roading operations are carried out to consistent high standards. Guidelines should consider the following phases before road construction starts:

- planning roads;
- design of roads;
- construction of roads;
- road use and maintenance;
- closure of roads.

1 Planning roads

Planning the road network for a forestry operation should be considered an integral part of the management planning process. Procedures for road planning should be drawn up. Pre-planning the

Box 16.1 Potential negative impacts of forest roading

- *Direct loss of forest cover* where roads are built. The more roads there are, and the wider each road is, the more forest has to be cleared to make way for them. This can result in significant losses of forest area.

- *Soil erosion* from the road surface and bare verges. Building roads involves clearing vegetation cover from the soil. Bare soil is more susceptible to erosion than when it is protected and held together by plants' roots. Soil erosion can lead to a loss in productivity of areas of forest near the road through the loss of nutrient-rich top soil.

- *Sedimentation of watercourses*, especially around bridges and fords. Soil eroded from roads is often carried into watercourses. This affects the water quality for aquatic animals and plants, as well as for people who may rely on the watercourse for drinking supplies. Sedimentation of watercourses is particularly severe where water runs directly from the road into the watercourse.

- *Access* for hunters, farmers, miners or illegal loggers. Forest roads not only provide access for forestry workers, but may also have the indirect impact of increasing hunting or agricultural pressures on the forest, or cultural impacts associated with immigration.

- *Disruption of animal migration patterns.* Roads may act as barriers to the dispersal and migration of forest-living animals. Animals which dwell in the forest canopy and which cannot cross from one tree crown to the next are at particular risk from roads.

- *Damage to the landscape.* In areas of high scenic and recreational value, roads can scar the landscape values associated with forests.

road network almost always results in a lower road density than ad hoc construction (see Box 16.2). Planning may also reduce road-building and maintenance costs and ensure better-located roads.
Planning procedures should include:

- *Defining maximum road density.* The higher the road density, the more forest area is destroyed.
- *Location of roads to minimize earthworks* and to facilitate drainage. Ridgetops are often favoured for this reason.
- Location of roads to *keep gradients to a minimum* and follow contours where possible.
- *Identification and protection of streamside buffer strips.* If a road has to follow a river valley, it must be located outside the streamside buffer strips.
- Route planning to *ensure the long-term stability of the soil and site.* Where possible avoid box cuts, which are difficult to drain without run-off directly entering watercourses. Minimize soil displacement, especially cut and fill, as it leaves unstable and easily eroded bare soil. Avoid side-casting on steep, erosive slopes, especially above watercourses.

Box 16.2 Road density

The density of forest roads refers to the area of land taken up by roads per hectare of forest. It is normally expressed as the number of metres (length) of road per hectare: knowing the average width of the road allows the area to be calculated per hectare. For example, a road density of 10 m/ha of roads which are 20 m wide means that 200 m^2/ha or 2 per cent of the ground will be covered by roads.

- Keep costs down and prevent environmental damage by *avoiding areas of wet soil or areas of high erosion risk*.
- Roads can *double as effective fire breaks* in plantations and fire-prone forests.
- *Avoid key biodiversity protection zones* and High Conservation Value Forests.

2 Design of roads

Specifications for road design should minimize negative impacts and improve safety. All employees involved in road design and construction should be familiar with the road design standards. Local or national standards may be adequate and must be observed, but in some cases they do not exist and need to be developed according to local conditions. Table 16.1 shows an example of road standards from the Philippines.

Specifications should include:

- *Maximum width for different categories of roads*, such as access, primary or secondary. All roads should be wide enough to allow safe passage of vehicles, but as narrow as possible to avoid excess forest clearing and exposure of soils.
- *Maximum clearing widths for roads*. Especially on clay and loam soils, trees need to be cleared back from the road shoulders enough to allow sunlight onto the road, to dry the surface after rain. This should be balanced with a need to minimize forest disturbance. Where wide clearings are essential, crossing points for canopy-dwelling animals should be left at intervals where trees are allowed to meet over the roadway – preferably over better-draining soils.
- *Maximum gradients and maximum sustained gradients*, including favourable (downhill when trucks loaded) and adverse (uphill when loaded) gradients, with a specified length of slope over which steeper gradients may be permitted.
- *Minimum camber* necessary to ensure water runs off the surface.
- *Minimum culvert* dimensions and frequency: culverts should be large and frequent enough to ensure all dips are covered and can carry peak water flows during heavy rains.
- *Culvert type*, for example wood, steel or concrete, depending on the durability required and the relative costs of the materials.

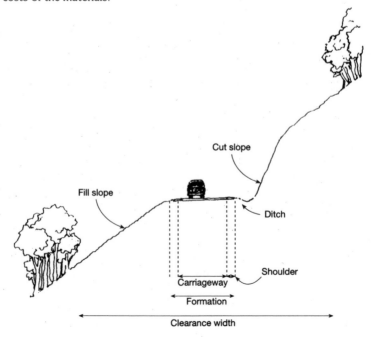

TABLE 16.1 Sample road standards from model medium-term Forest Management Plan for Suriago Development Corporation 1993–2002, Philippines

Item	Primary Roads	Secondary Roads	Skid Trails
Practicability	permanent	temporary	seasonal
Road surface	gravel	gravel	earth
Number of lanes	2	1	1
Maximum distance between turnout drains	–	200 m	–
Minimum width			3–4 m
Formation	8 m	6 m	
Carriageway	6 m	4 m	
Minimum curve radius	50 m	30 m	–
Maximum sustained grade			
Favourable	8%	10%	25%
Adverse	6%	8%	15%
Maximum grade			
Favourable	10%	12%	35%
Adverse	8%	10%	20%
Maximum cut-slope ratio			
Solid rock	4:1	4:1	–
Other materials	1:1	1:1	–
Maximum fill-slope ratio	1:1.5	1:1.5	–
Minimum ditch			
Width	1.0 m	1.0 m	–
Depth	0.5 m	0.5 m	–

- *Minimum side drain dimensions* to carry peak rain flows.
- *Frequency of turnout drains* carrying water out of the side drains to disperse in the forest. Frequency depends on the amount of water, slope and soil type.
- *Maximum length of box cuts permitted.* Longer and steeper box cuts channel water faster, causing more erosion of the road surface and increased sedimentation of waterways at the bottom.
- *Maximum gradient of sidecut slopes,* for both cut-slope and fill-slope ratios.
- *Minimum visibility distance* around bends, especially where two trucks may meet at high speeds.

3 Construction of roads

During road construction large amounts of loose soil are moved around by heavy machinery. Bridge construction may involve working in the watercourse, potentially damaging the streambed and banks. Soil erosion and damage to watercourses may be severe. To minimize impacts during construction and to ensure the specifications are met, operations should be carried out by a trained roads crew under careful supervision. Some important issues to consider in construction procedures are:

- *Timing* of road construction to be during the dry season if possible (in seasonal climates) to prevent excessive run-off and erosion. Road compaction should be completed before the rains begin, and a minimum time period specified for consolidation before use.
- *Prohibition or minimization of heavy machinery use* in watercourses, during construction of bridges, to prevent damage to the streambed.
- *Minimization or prohibition of cut and fill.* Where cut and fill is unavoidable, side-cast material should not be cast downhill into a watercourse. Revegetation of fill slopes as soon as possible after construction to hold together the loose soil.
- *Construction of bridges and culverts* to avoid changing water flow patterns.

4 Road use and maintenance

Guidelines covering road use and maintenance should be developed. Issues to consider are:

- *Definition of who has rights* to use the road, if the forest organization can restrict access.
- *Safety* of employees and other road users. Speed limits and warnings at dangerous bends may be necessary.
- *Warning signs* that logging trucks are using the road, if public access is permitted.
- *Restrictions on road use* during and after heavy rain, particularly in areas of high rainfall. For example heavy trucks passing along a wet, unsealed road will damage the road and add to maintenance costs.
- *Schedule of routine checks* and maintenance of roads, particularly of drainage structures.

5 Road closure

Once built, roads should generally be considered as permanent structures as their compacted surface will not easily revegetate and they are no longer part of the production area. Where possible, it should be planned to use the same roads again for later cutting cycles. In the intervening time, access to the road should be controlled if possible.

Consultation with local stakeholders should aim to find out if a road is required for uses compatible with sustainable forest management. If it is, then responsibilities for maintenance and rehabilitation need to be agreed by the stakeholders and the forest organization

Then, if closure is required this might entail:

- *Removal of bridges* and culverts to prevent road use;
- *Clearance of watercourses* to ensure they are flowing freely;
- *Installation of water bars* to prevent water from running along the road surface.

Water bars prevent water from running along the road surface

16.2 Harvesting and Extraction

WHY IS HARVESTING AND EXTRACTION IMPORTANT?

Harvesting and extraction operations are the activities which generally cause the most significant impacts on the forest. They include all the activities necessary to fell trees and remove them from the forest to the log landing, roadside or other sites for loading and transport. The quality of planning and execution of harvesting and extraction is crucial in determining the state of the forest ecosystem following harvesting.

Poorly executed harvesting and extraction can damage the residual stand and site, to the extent that future harvests are not possible. Some potential negative impacts of harvesting and extraction are listed in Box 16.3. In contrast, reduced impact logging can also lead to cost savings (see Box 16.4).

WHAT IS REQUIRED?

In order to ensure that harvesting and extraction operations are carried out to the highest standards, documented procedures should be developed and implemented. These may be based on the FAO Model Code of Forest Harvesting Practice[1] and should incorporate appropriate local standards or harvesting standards.

Box 16.3	**Potential impacts of harvesting and extraction on the forest ecosystem**

Harvesting and extraction operations have a number of major impacts:

1   *Impacts on soils.* Soil disturbance is caused particularly by ground skidding and high-loading, resulting in soil compaction and erosion. Nutrient loss may occur through felling and extraction. Soil compaction caused by the passage of heavy machinery reduces the soil's water absorption capacity. Water then runs over the top, causing soil erosion. Plants' roots grow poorly in the anaerobic conditions of compacted soils, affecting regrowth and jeopardizing future timber yields. Soil nutrient loss is more likely with whole-tree removals and fast-growing species through consistent removal of nutrients from the system.

2   *Impacts on water.* Extraction of logs by dragging them across waterways can damage the streambed and interfere with the water flow. increased levels of soil erosion on tracks and trails cause increased levels of sediment in waterways, which in turn affects the water quality for aquatic animals and plants. Serious sedimentation of waterways can also affect reservoirs and dams downstream.

3   *Impacts on the residual stand.* In selection systems, felled trees damage and break other trees as they fall. Vines (lianas), tying tree crowns together, can pull other trees over. Extraction machinery and logs being extracted break tree stems and branches, scrape bark off mature trees, and crush smaller trees and seedlings. Typically, for every tree which is logged in many moist tropical forests, a second is destroyed and a third is damaged beyond recovery. This may affect the ability of the forest to regenerate and therefore the prospects for future harvests. It is also avoidable.[2]

4.   *Other impacts.* For example, the impact on wildlife and disruption to breeding and nesting as well as landscape damage, particularly from large clear-felling in visually prominent areas.

---

1 DP Dykstra and R Heinrich, FAO Model Code of Forest Harvesting Practice. Food and Agriculture Organization of the United Nations, 1996. ISBN 92-5-103690-X

2 Bruijnzeel and Critchley, *Environmental Impacts of Logging Moist Tropical Forests*, IHP Humid Tropics Programme Series No 7, Division of Water Science, UNESCO, 1994 (See contact details for UNESCO, Appendix 4.2)

| Box 16.4 | **Costs and benefits of implementing a reduced impact logging system** |

1   *The CELOS Harvesting System, Surinam:* The increased costs of planning have been found to pay off in increased efficiency and reduced wastage, as well as reduced environmental damage. Well-planned and controlled harvesting and extraction operations were found to reduce total logging costs by 16–31 per cent compared to traditional logging methods. The cost savings were made largely through improved efficiency, especially skidder use.[3]

2   *Strategies for sustainable wood industries in Sarawak (ITTO Project):* Initial cost estimations suggested savings in skidding costs of approximately US$20 per hectare if pre-planned skid trails were used, with greater savings if the length of skid trails was reduced. However, these savings have to be offset against increased planning, engineering and supervision costs.[4]

3   *Tropical Forest Foundation, Fazenda Cauaxi, South-west Paragominas, Pará, Brazil:* Initial harvest levels were approximately 25 m³ per hectare. Wood wasted in conventional logging (CL) blocks represented about 24 per cent of initial harvest volume, compared to 8 per cent in reduced impact logging (RIL) operation. More careful bucking, better coordination between felling and skidding crews and better tree selection improved volume recovery. Planning and infrastructure activities increased 'upfront' costs by about 170 per cent over CL operations. Felling and bucking costs were also higher. However, efficiency gains from RIL were large. Overall, costs per cubic metre associated with a typical RIL system were estimated to be 12 per cent lower than the costs of a typical CL system.[5]

Procedures should cover the following areas:

- harvest planning;
- tree marking;
- felling and extraction;
- post-harvest operations.

1 Harvest planning

Advance planning of harvest operations can significantly reduce both costs and damage to the residual stand or site. The details required are determined by the silvicultural system, as well as the local conditions of climate, topography and forest types. The following steps should be considered:

- *Planning of harvesting operations:* locations and sequences of areas to be harvested should be determined.
- *Pre-harvest inventory:* the design depends on the type of forest and information needed. For selection systems in natural forest, this should involve a stock survey of commercial trees (100 per cent enumeration: see Box 15.11). For clear-cut systems, it may involve an inventory to determine likely harvest levels.
- *Analysis of inventory data* and production of a stock map where appropriate (see Box 16.5).
- *Identification and demarcation* of areas to be avoided during harvesting and extraction, especially streamside buffer zones, steep slopes, fragile soils, wet ground, cultural heritage sites and biodiversity protection zones.

3 Bruijnzeel and Critchley, *Environmental Impacts of Logging Moist Tropical Forests.* UNESCO, 1994

4 Putz & Pinard, 'Reduced impact, increased costs?' ITTO *Tropical Forest Update,* Volume 6, No 3, 1996

5 TP Holmes, GM Blate, JC Zweede et al, *Financial Costs and Benefits of Reduced-Impact Logging in the Eastern Amazon.* Tropical Forest Foundation, Washington, DC, 2000

Box 16.5   **Pre-harvest planning: the stock map**

The results of a pre-harvest stock survey are often presented as a stock map, which is used for detailed harvest planning:

- which trees to harvest;

- what species and sizes;

- which direction to fell them;

- extraction routes.

A sample of a stock map is shown below; this can be hand drawn or produced on a computer:

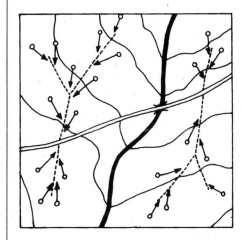

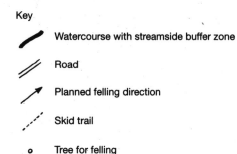

Key

/   Watercourse with streamside buffer zone

//   Road

↗   Planned felling direction

⋯   Skid trail

∘   Tree for felling

- *Planning of extraction methods and routes* including ground skidding and skyline systems. Reduced impact extraction methods include planning and marking extraction routes prior to felling and winching logs from stump to skid trail in ground skidding systems, which avoids damage to the residual stand.
- *Planning of maximum clear-cut size* (if clear-cut systems are being used): large clear-cuts are more prone to soil erosion and nutrient losses and are more visually disruptive than small ones. In general the size of clear-cut should be similar to the size of clearing made by natural disturbances, such as fires and storms. However, aesthetic considerations may also be important. Many countries have developed limits on the maximum size of clear-cut permitted: for example, 5–20 ha in Sweden; 4 ha in Poland.

2 Tree marking

Issues which should be covered in the procedures for tree marking and felling include:

- *Criteria for determining which trees should be felled, and which should be retained* as habitat trees, future crop trees, seed trees and for the benefit of keystone species (see Box 6.1 and Figure 16.1).
- *Criteria for deciding direction of fall* to minimize waste and breakage, damage to the residual stand and danger for the chainsaw operator (see Box 16.7).
- *Definition of responsibilities for tree marking*: for example the logging supervisor or tree marker.
- *Training*: ensure tree markers have sufficient training;

Box 16.6   **Addressing common problems with pre-harvest planning**

Pre-harvest planning is not providing useful information, not being used and/or not preventing harvesting and extraction damage.

Consider trying the following:

1   Ensure adequate and regular supervision of the quality of pre-harvest plans.

2   Where problems persist, bring together all the people involved in planning and execution of harvesting at different stages. Visit a harvested site where problems have occurred. Encourage everyone to discuss what has happened and why – though avoid personal allocation of blame. Agree what changes need to be made at what point in the process and develop procedures (written or otherwise) to address the issues.

3   Review the incentives being offered to harvesting and extraction crews to see how thay can be improved. Production incentives (eg payment per cubic metre felled) may encourage harvesting crews to ignore plans and fell easily accessible trees.

- *Methods for marking trees* to be felled and retained, and method of indicating intended direction of fall (eg paint marks, arrows, tags, etc)
- *Responsibility for cutting vines* (lianas) which may tie crowns of trees due to be felled to other trees, creating a safety risk and increasing felling damage. Ideally these should be cut several months in advance of felling (for example, during pre-harvest inventory or tree marking) to allow them time to dry out and become brittle.

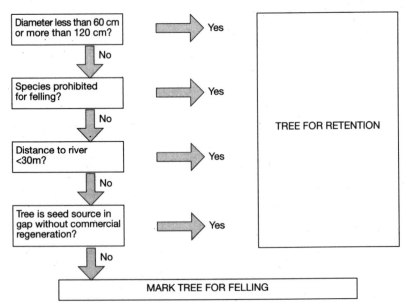

FIGURE 16.1 Tree-marking decision chart for timber harvesting[6]

---

6 Adapted from M Kleine and D Chai *Tree Marking in Logged-Over Dipterocarp Forests*. Sabah Forest Department, Malaysia, 1994

Box 16.7 **Deciding direction of fall[7]**

Directional felling includes a variety of techniques with a chainsaw for controlling the direction in which a tree falls. Directional felling requires good training and constant supervision, as well as appropriate and well-maintained equipment, and appropriate incentives for the chainsaw operator. The direction of fall should be decided according to the following criteria:

- *Extraction direction:* Logs are easier to extract, and cause less damage, when felled at an angle (30–45° is ideal) towards the extraction route.

- *Minimal damage to the log and to the residual stand:* Where possible, trees should be felled into existing canopy gaps or on to a skid trail. To cause minimal damage, the crown should be felled on top of the crown of a previously felled tree, reducing the area affected and cushioning the impact. The felling direction should also avoid hitting potential crop trees and young regeneration of commercial species.

- *Protecting seed trees:* Some trees of good form of commercial species should be left standing in order to provide seeds for the next generation of crop trees. These trees should be marked to assist the feller to identify them and to reinforce the message that they should not be damaged.

- *Steep slopes:* Very steep slopes should be avoided in harvesting. Where felling is done on slopes, trees should be felled across the slope where possible, to prevent breakage of the log and minimize its slide downhill. This is not always possible, if the lean of the tree is strongly downhill.

- *Feller's safety:* The direction of fall should always allow the feller a safe exit route. Trees should not be felled if they will require a balancing act from the feller.

3 Felling and extraction

Procedures should include:

- *Felling rules and techniques to be used,* for example the number of men in a crew, safety measures, methods of checking for rot, directional felling techniques and emergency procedures;
- *Opening extraction routes prior to felling.* Particularly in the case of ground skidding in selection systems, marking and opening of main skid trails prior to commencement of felling reduces the area of skid trails created and therefore the amount of damage to the residual stand. Skid trails should be planned using the stock map, to ensure the most efficient route.
- *Extraction methods, particularly from stump to skid trail in ground skidding systems.* Using a winch, rather than approaching each stump with a skidder, reduces damage to the residual stand.
- *Extraction to landing* and handling at landing, including storage, loading and unloading.
- *Methods of numbering and recording* harvested and extracted trees to enable tracking from stump to processing. This helps to ensure that all trees identified for harvest are felled, extracted and trucked, minimizing abandoned logs and waste in the forest. It is also essential for harvest monitoring. Increasingly, processing companies need to demonstrate that their raw materials come from legal sources. Robust methods of identifying and recording harvested trees, preferably linked to pre-harvest stock survey information, can provide these assurances.
- *Training:* experience from many countries suggests that many forest operations will require a significant improvement in worker training if they are to implement guidelines for better felling and extraction.

---

7 See: DP Dykstra and R Heinrich, FAO *Model Code of Forest Harvesting Practice.* Food and Agriculture Organization of the United Nations, Rome, 1996

> **Box 16.8** **Reduced impact logging – minimizing damage to the residual stand**
>
> Experiments with the CELOS (reduced impact) harvesting system in Suriname reduced damage to the remaining stand by up to 40 per cent through improved planning and execution of felling and extraction. Felling eight to ten trees per hectare, approximately 13 per cent of the area was damaged by felling gaps in a traditional operation, compared to 7 per cent in controlled felling. Similarly, skidding damage was reduced from 15 per cent to 7 per cent by better control of extraction. Use of the winch to extract logs from stump to skid trail further reduced damage to the remaining forest to 5 per cent.[8]

4 Post-harvest operations

Issues to consider following harvesting and extraction include:

- *Post-harvest assessments* of site condition (see Box 17.1) to ensure harvesting and extraction were carried out according to procedures.
- *Post-harvest site rehabilitation* methods, such as ripping and replanting of landing sites and skid trails; construction of water bars on sloping skid trails to prevent erosion; distribution of brash and harvest residues.
- *Methods of block closure*, such as removal of bridges and culverts, and means of enforcement of the minimum time period before re-entry to allow the forest sufficient time to recover. Conditions under which re-entry is permitted, before the minimum time period, should be specified, for example for silvicultural treatments.

16.3 Silviculture

WHY IS SILVICULTURE IMPORTANT?

Silviculture is the art and science of growing trees. It involves manipulating natural biological processes of the forest in order to achieve specific end results. It includes all operations that are done between one harvest and the next, such as planting, thinning, pruning, weeding or poison girdling. Harvesting operations themselves are a major silvicultural treatment.

The choice of silvicultural system is crucial in determining whether the forest can produce the desired products and services on a sustainable basis. The sustainability of the forest ecosystem under management depends on the silvicultural systems applied – applying the wrong silvicultural system will damage the forest's ability to regenerate or provide desired results.

WHAT IS REQUIRED?

Some areas of the world and some forest types have well-established silvicultural systems; in other areas the system is still being developed. A silvicultural system should be based on:

---

8 J Hendrison, 'Damage-controlled logging in managed tropical rain forest in Suriname', *Ecology and Management of Tropical Rain Forests in Suriname*, 4) Wageningen Agricultural University, The Netherlands, 1990. Cited in: H ter Steege et al, *Ecology and Logging in a Tropical Rain Forest in Guyana*, Tropenbos Series 14, 1996, Stichting Tropenbos. ISBN 90-5113-026-0

Box 16.9   Addressing common problems with silvicultural systems

For many areas of the world, adequate silvicultural systems are not well defined. Because silvicultural systems depend on so many factors they cannot be taken wholesale from one region to another. A system which works in one forest type may not work in another. Long-term research is generally needed to understand the effects of individual silvicultural interventions and their combined effects.

In the absence of a well-defined silvicultural system, basic steps the forest manager may consider are:

1   Look at *management in similar forest types:* in particular, harvesting practices can be a silvicultural tool if well planned and executed (see Section 16.2).

2   Consider the *appropriate harvest intensity*, taking into consideration the damage caused by higher harvest intensities and the need to leave potential crop trees and seed trees for the next felling cycle (see Section 15.5).

3   *Establish permanent sample plots* to monitor the effects of current management practices and provide a baseline for comparison with growth rates following experimental silvicultural treatments (see Section 17.2).

4   *Keep records* of current harvest and damage levels to allow the long-term effects of harvest intensities to be understood.

5   *Support on-going research* to define appropriate post-harvest silvicultural treatments.

- the forest type;
- current species composition;
- environmental conditions, such as soil, topography and climate;
- labour and inputs available for management;
- the end products and services required.

Some of the possible effects of the choice of silvicultural system are described below. These should be considered when developing or implementing a silvicultural system and should form part of the description and justification of the silvicultural system in the forest management plan.

Issues to consider when developing a silvicultural system include:

- effects on growth rates of harvestable species (timber and non-timber);
- effects on forest services;
- effects on biodiversity and wildlife;
- effects on High Conservation Values;
- effects on ecological sustainability;
- effects on seed production;
- effects on regeneration;
- applicability to forest ecology;
- social acceptability.

1 Effects on growth rates of harvestable species

Growth rates are affected by the intensity of harvest, the species harvested and silvicultural interventions such as fertilization, poison girdling or thinning. For example, experiments in natural forest in Côte d'Ivoire showed that thinning non-marketable species resulted in an increase of 50–100 per cent in diameter growth of small to medium-sized marketable stems.[9]

2 Effects on forest services
Silviculture can affect forest services such as watershed protection, recreation or aesthetic value. For example, in Poland, around 15 different silvicultural systems are used depending on the terrain and forest type, with a particular emphasis on maintaining these services. Steep slopes are harvested lightly and extraction is carried out using horses in order to maintain the protective functions of the forest. On lesser slopes more intensive silvicultural systems may be applied.

3 Effects on biodiversity and wildlife
Because intensive silviculture simplifies the structure of the forest, it reduces biodiversity. In such systems, protection of unharvested areas for conservation of biodiversity should be considered (see section 15.6). For example, in the CELOS system, developed in Surinam, natural forest was 'refined' by the removal of unmarketable species, simplifying the forest ecosystem. To compensate, up to 10 per cent of the forest area was designated to be left unharvested for conservation of biodiversity.[10]

4 Effects on High Conservation Values
Where High Conservation Values (HCVs) have been identified in the forest, silvicultural systems should be designed to maintain or enhance the HCVs. This may mean deliberately promoting some types of habitat, for example by leaving dead wood habitats where they are critical to the survival of a range of rare, threatened and endangered species.

5 Effects on ecological sustainability
The silvicultural system will affect ecological processes such as nutrient and water cycles. For example, afforestation with fast-growing trees has been blamed for reduced dry-season water flows in South Africa, because trees lose more water to the atmosphere than the original grassland.[11] Nutrient losses from the ecosystem are affected by the harvesting system (clear-felling or selection), by the gap size left by felling, by the soil type and by the proportion of the tree (including its bark and leaves) which is removed from the site, all of which are part of the silvicultural system.

6 Effects on seed production
Where natural regeneration is used, seeds for the next generation of trees are provided by the existing trees. Consideration should be given to leaving sufficient seed trees to ensure trees of good form continue to regenerate. The silvicultural system should ensure that:

- trees of good form are left at each harvest;
- a minimum number of seed trees per hectare is defined;
- cutting limits take account of the size when seed is produced;
- seed trees are accessible to pollinators and dispersers, and not isolated from the rest of the forest.

For example, in Costa Rica, CODEFORSA a private forest management company working with owners of small blocks of natural forest, ensures that 40 per cent of harvestable stems are left during the first harvest to provide a seed source and as the basis of the second harvest.

---

9 H F Maitre, 'Silvicultural interventions and their effects on forest dynamics and production in some rain forests of Cote d'Ivoire', in eds. A Gomez-Pompa, T C Whitmore & M Hadley *Rain Forest Regeneration and Management* 1991

10 NR de Graff, 'A silvicultural system for natural regeneration of tropical rain forest in Suriname. *Ecology and Management of Tropical Rain Forests in Suriname* 1) Wageningen Agricultural University, Wageningen, The Netherlands, 1986

11 *Benefits and Costs of Plantation Forestry: Case Studies from Mpumalanga*, Department of Water Affairs and Forestry, South Africa, 1995

7 Effects on regeneration

Where natural regeneration is used, the type of harvesting system employed and silvicultural treatments employed will affect regeneration. Often silvicultural systems for selective harvesting are based on simple cutting limits, such as minimum diameters for harvestable trees. This does not consider the effects on regeneration. For example, where the predominant timber species regenerate well in light conditions, a silvicultural system based on high cutting limits may not be successful. Removing trees of large diameter only may not allow enough light through to encourage regeneration.

8 Applicability to forest ecology

Some forest ecosystems are adapted to periodic disturbances of small to large scale, such as fire or hurricane damage. The silvicultural system should mimic these disturbances to create the right conditions for growth and regeneration.

9 Social acceptability

The silvicultural system needs to be socially acceptable, particularly to surrounding communities. For example, in some instances clear-cutting might be silviculturally appropriate, but because it is highly visible and because many people do not like to see clear-cuts – a compromise must then be reached. Some silvicultural systems allow for grazing of cattle between trees, or inter-cropping plantation trees with crops.

All the above effects may be intensified or reduced by the way in which silvicultural operations are carried out. Well-executed and supervised silvicultural operations are essential to ensure that the end effects are as predicted. Setting a harvest limit of five trees per hectare for a selection system, but destroying an extra ten trees during harvesting, will not produce the desired results (see Section 16.2). The silvicultural system needs to be closely linked to guidelines and procedures for carrying out other activities.

16.4 Chemicals and Pest Management

WHY IS CHEMICAL AND PEST MANAGEMENT IMPORTANT?

Chemicals used by forest managers include a range of fertilizers, herbicides, insecticides, fungicides and hormones. Insecticides and fungicides are both pesticides and can be used on animals or plants which cause damage to trees or any other useful plant. Chemicals are most commonly applied in nurseries and young plantations. Although rates of chemical usage are much lower than in agriculture, over the life of the crop, its application tends to be concentrated in short periods.

Few chemicals are completely specific to their target, as a result their effects can be widespread and long term (see Box 16.10).

WHAT IS REQUIRED?

Procedures and guidelines should be developed and implemented to cover the following stages:

1 planning and procurement of chemicals;
2 storage of chemicals;
3 use of chemicals;
4 disposal of chemicals and their containers.

Some countries have legislation which covers all four stages and applicators may also need to be licensed.

Box 16.10 **Some potential negative effects of chemicals in forestry use**

Pesticides may have some of the following adverse effects:

- Non-target species in the forest may be affected.

- Chemicals may drift on to nearby crops.

- Animals and plants in nearby watercourses may be affected.

- Drinking water quality may be affected.

- People living nearby and operators applying chemicals may be harmed.

- Biological diversity may be reduced.

- They may enter the food chain, affecting wildlife and livestock seemingly unconnected with the pests or chemical application.

Fertilizers can wash off into watercourses, affecting water quality and potentially 'killing' lakes and standing water bodies. This simple flow chart highlights the stages in the death of a lake as a result of excess fertilizer application at a plantation.

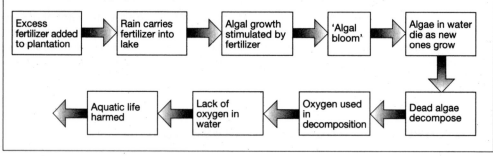

---

1 Planning and procurement of chemicals

Planning guidelines should consider both the need for chemicals in forest management and means of minimizing their use. Where chemical use is essential, the process of application must be also be planned.

Integrated pest management combines preventative and remedial procedures for controlling pests. It is a possible means of reducing the pesticide inputs needed for forest management by using good silvicultural techniques and management practices (see Box 16.11).

Integrated pest management needs forward planning to be successful and it requires detailed knowledge about the ecology of the pest and its host tree. It also tends to be labour intensive and as no two situations are ever the same, pest management techniques which are successful in one place may not work in another. Specialist help is therefore need in its application. It is generally suited only to plantation forestry.

If chemicals are considered necessary for forest operations, the following aspects should be considered in planning their use and incorporated into procedures:

Which chemicals will be used for which jobs?
- Define objectives of chemical use and consider any alternative options.
- Ensure the chemicals chosen are appropriate to the job and keep up to date on the safest and least environmentally damaging chemicals. Manufacturers are under pressure to improve performance in these respects and are producing new materials.

Box 16.11 **Summary of procedures available for the integrated pest management of forest insects[12]**

Integrated pest management is a combination of preventative and remedial procedures for controlling forest pests. It is encouraged because it relies as much on preventing problems in the first place by good forestry practices, as on combating problems once they have taken hold.

Suggested order of priority | *Methods for preventing or minimizing pest problems:*

1   Choose appropriate site, tree species or provenance; select suitable origin of tree (native or exotic), planting regime (mixtures or monocultures).

2   Use good silvicultural practices (sanitation, hygiene, maintenance of tree vigour).

3   Establish quarantine measures against exotic pests; define internal quarantines (within the forest management unit) for established ones.

4   Survey for actual or potential pests; assess size and impact of problem; routinely monitor serious pest populations (life stage counts, pheromone traps).

5   Make management decisions based on all available information related to economic, entomological and silvicultural factors.

       Methods for controlling pest outbreaks:

6   Silvicultural or ecological techniques (selective felling, sanitation and hygiene felling, trap trees).

7   Biological control with pathogens, parasites or predators where appropriate.

8   Insecticidal control using selective chemicals applied efficiently.

- Ensure that none of these chemicals are on the WHO lists 1a and 1b (see Appendix 3.1) or contain chlorinated hydrocarbons.
- Maintain a list of all chemicals, their uses and reasons for their use. Keep records of quantities used and locations of application.
- Ensure that chemicals proposed for use are not included in WHO Classes 1a and 1b, are not chlorinated hydrocarbons or other persistent, toxic or accumulative pesticides. Chemicals prohibited under FSC-based standards are listed in Appendix 3.1.

Where will chemicals be used?

- Define situations and locations in which a chemical will be used.
- Prohibit chemical use in environmentally sensitive areas (such as streamside buffer strips and conservation areas).
- Prohibit chemical use in areas where there is a danger of them being carried into watercourses or filtering into groundwater.

When will chemicals be used?

- Guidelines for efficient timing of chemical application. For example, plan fertilizer applications to coincide with the growing season of target plants.
- Rules for timing of pesticide applications. For example, prohibit use during heavy rains, or when heavy rain is expected. Do not spray in windy conditions.

Who is responsible for chemical use?
- Appoint a member of the management team to be responsible for chemical use.
- Restrict access to chemicals to authorized personnel.
- Limit times at which chemicals are distributed.

How will chemicals be used?
- Considerations of the operator's health. For example, by providing protective clothing and training.
- Application methods to maximize efficiency and minimize environmental contamination.

What should be done in the event of an accident or spillage?
- Contingency plans for medical or environmental effects of accidents.
- Identify personnel responsible for containment and the methods which should be used for clean-up operations.

2 Storage of chemicals
Chemical storage areas should be secure. Store rooms must meet national standards. Guidelines for chemical storage should consider:

Where are chemicals stored?
- Storage should generally be in secure, dry, cool buildings.
- Means of restricting access to authorized personnel and clear locking procedures.

Containers for transport and storage
- Clearly labelled containers, with the chemical name and a warning if it is hazardous.
- Storage off the ground to allow easy identification of leaks.
- Store minimum amounts of chemicals in the field, preferably only one day's supply.

| Box 16.12 | **Addressing common problems with chemicals** |

Protective clothing recommended by chemical companies is frequently designed for use in temperate areas and is uncomfortable in tropical conditions. Operators using chemicals do not use the protective clothing provided because of the discomfort.

Consider trying to:

1 Contact the chemical supplier (and universities/research institutions) to find out what the most important health risks to operators are: from contact with skin, breathing chemical fumes, accidental spillages, etc.

2 Discuss the problem with suppliers and point out to them that operators may suffer from health problems associated with the protective clothing or equipment.

3 Work with the suppliers to prioritize the health risks and determine what is the minimum equipment which will adequately protect operators and whether more appropriate equipment can be designed.

Control of stores
- Maintain inventory of all chemicals kept in storage and their dates of purchase.
- Control issue of chemicals, ensuring oldest batches are used first.

3 Use of chemicals
Guidelines for use of all chemicals should be developed. Local health and safety regulations should be taken into account. Monitoring and unannounced spot checks assist in ensuring that plans and regulations are implemented (see Chapter 13). Issues to be considered by guidelines and procedures for chemical use include:

Training
Operators should know:

- what they are using;
- why they are using it;
- how they should prepare and use it;
- what the potential dangers are;
- what to do if something goes wrong;
- basic first aid techniques associated with chemical use.

Equipment
- Appropriate protective clothing should be available and in use as described in the chemical data sheet.
- Appropriate and properly functioning equipment should be used.
- Facilities should be available for chemical preparation and for cleaning equipment after use.
- Equipment must be available to deal with accidental spillages as set out in the contingency plans.

Environmental protection
- Ensure operators are aware of and respect areas where chemicals are not to be used.
- Ensure operators know techniques for minimizing the amount of chemical used, such as spot applications.
- Ensure operators do not dispose of chemicals, or wash equipment, in watercourses.

4 Disposal of chemicals and their containers

Guidelines should include:

Disposal of chemicals:
- Dispose of excess and outdated chemicals.
- Designate location of disposal sites away from watercourses, according to local regulations.
- Define procedures for chemical disposal.

Disposal of chemical containers
- Negotiate with suppliers or manufacturers to take back empty containers.
- Ensure empty containers are not re-used for other purposes (especially for storage of food and water).

16.5 Training

WHY IS TRAINING IMPORTANT?

Even with the best planning and equipment, forest organizations cannot achieve or maintain SFM if the people carrying out operations do not have the necessary skills or information. At all levels, training is needed to ensure that operations are carried out according to required procedures and to maintain high standards. Training can be formal or informal, on-the-job or in-class, evening classes or workshops, on-off or on-going (Figure 16.2). Some of the potential benefits of training are shown in Box 16.13.

On the job

Workshop

Informal

Formal

FIGURE 16.2 **Types of training**

WHAT IS REQUIRED?

Training should be an on-going process of:

- planning and implementation;
- evaluation and feedback.

Although the type of training programme outlined below is more appropriate to a large-scale organization which is planning its training provision for employees, the same process can be used to determine training needs and possibilities for small organizations and community groups. It is important to incorporate provision of training into on-going staff management and development programmes.

The essential features are to ensure that one person is appointed as responsible for training; that training is made relevant to the individuals' and organization's needs and that there are mechanisms for feedback into improved training provision.

1 Planning and implementation
- *Appoint a member of management* with responsibility, authority and a budget for training.
- *Review or establish job descriptions* for all operators or employees.
- *Establish which skills or qualifications are necessary* to carry out those jobs effectively, remembering that new skills and working disciplines are necessary for the implementation of sustainable forest management. It can be tempting for senior management, in particular, to ignore the potential of skills which they do not personally possess.
- *Compare the skills and qualification needed with those available*, to determine what training is needed.
- Discuss with employees or operators their perceived training needs to carry out their responsibilities satisfactorily.
- *Prioritize training needs* depending on factors such as:
  - *The severity of impacts caused by the lack of training*: For example, poor felling practice is dangerous for chainsaw operators, so training in felling may be considered a priority because safety is paramount.

Box 16.13 **Potential benefits of training**

- Operators and managers possess the knowledge and skills needed to carry out their jobs effectively.

- Better skills improve performance and build confidence – both are essential for a continuous improvement programme.

- Training motivates people and is important in retaining commitment and drive.

- Well-trained people can work more flexibly and may be able to contribute in a greater variety of ways than previously.

- As roles and responsibilities of staff change, training ensures they can carry out their new roles satisfactorily.

- Training opportunities within a job make it attractive to high quality candidates.

- Training reduces waste and may lower costs.

- Training should enhance the earning potential of operators.

Box 16.14 **Addressing common problems with training**

Training is ineffective and results are not incorporated into operations.
  Consider trying the following:

1   Focus on determining what are the real needs for training at all levels and target training resources efficiently.

2   Make sure that adequate emphasis is place on operational level training and that it is given equal importance with professional/managerial level training.

3   If training courses – particularly operational or practical courses – are not available, try working with other companies to bring in external trainers.

4   Contact training institutes (such as the Centre for Rural Development and Training) for advice on finding appropriate trainers and courses.

5   Try to build internal training capacity, particularly in providing operational level, on-the-job training, but make sure the trainers are regularly updated themselves.

6   Make sure people are given the opportunity to use their training in their jobs: don't train people in subjects they cannot put into practice. They'll only forget it!

- *Organizational policy and objectives*: For example, if reduced damage from skidding operations is an objective of the organization, training for skidder operators and supervisors who mark skid trails could be a priority.
- *Funds and resources available*: For example, sponsorship for long-term university courses may need to be weighed up against benefits of short-term, on-the-job training.
- *Consider the most appropriate ways to provide training.* For example, on-the-job training, formal/informal training, public training courses, internally organized training, evening classes, part-time university study workshops, and so on.
- *Compile a schedule or programme* of training to be provided and discuss with potential trainees to ensure their support.
- During implementation, *keep records* of training needs analysis, training proposed and training received by all personnel.

2 Evaluation and feedback on training

Evaluation and feedback should assess whether the aims of training were met and how training could be improved. Consideration should be given to:

- *Regular reviews of the training programme.* Every six months hold a meeting with senior management to review training and the improvements it has made in staff's skill levels and in helping the organization meet its objectives. Adapt the training programme accordingly.
- *Provide a means for feedback from trainees.* Questionnaires are one way to allow staff and trainers to assess the usefulness of training attended. Alternatively, discussions with trainees on the job or debriefing sessions with the training manager after training is completed can ensure better and better training programmes.
- *Establish feedback links with internal audit* (see Chapter 13) to give an objective measure of improvements in day-to-day operations.
- *Ensure trained staff are given the opportunities to exercise new skills* in the course of their day-to-day work. If they are encouraged to do this, the training programme will lead to improvements and changes in forest management.

17 Monitoring

Monitoring is the process of collecting information about how effectively management prescriptions are being implemented, and what effects those prescriptions are having. Two levels of monitoring activities – operational and strategic monitoring – are described separately here.

Detailed and extensive monitoring can be expensive. The intensity of monitoring which should be carried out by an organization will depend on its size, the complexity of operations and the performance standards, as well as the resources available. Operational monitoring, such as audit activities, which monitor whether operations are being carried out according to plan, should be a normal part of management. Longer-term strategic monitoring may need specialist inputs and may not be feasible for smaller forest organizations.

17.1 Operational Monitoring

WHY IS OPERATIONAL MONITORING IMPORTANT?

Operational monitoring allows forest managers to identify whether practices are being properly implemented, or where the effects of certain operations are not as expected. Operational monitoring is important because it:

- *Identifies successes and failings of forest management practices* (for example, a post-harvest assessment might discover that skid trails frequently cross watercourses and damage stream beds – in contravention of operational procedures).
- *Provides a basis for modifications to future management* (for example, develop new guidelines to ensure that streamside buffer zones are demarcated in the field before felling commences).
- *Identifies areas where corrective (or remedial) action needs to be taken* (for example, ensure that watercourses blocked or damaged by skidding are cleared and restored).

WHAT IS REQUIRED?

Operational monitoring may comprise effective supervision of activities with informal record-keeping, regular management visits, and/or a more formal audit programme. (Internal audit has a specific meaning in terms of EMS – see Chapter 13.) An audit programme involves a regular and systematic evaluation of forest management performance against the standards laid out in procedures or the management plan. It may cover any management activity whose effectiveness the forest manager needs to assess. In a forest organization, these activities often include:

- harvesting;
- road construction and maintenance;
- silvicultural activities;
- health and safety;
- contractor activities;
- timber production.

Box 17.1   **Sample checklist for internal audit of harvesting site of tropical natural forest**

IDENTIFICATION:

Compartment name:

Block number:

Date:

Audit team:

Issue to check	*Yes/no*	*Observations*
1 Is the stock map/harvesting plan available and being properly used in the field?		
2 Are all trees for felling marked beforehand?		
3 Are chainsaw operators using directional felling effectively?		
4 Are skid trails marked in the forest prior to felling?		
5 Are stream buffer zones marked according to company guidelines?		
6 Do skidder drivers avoid stream buffer zones?		
7 Is all appropriate equipment available and in working order?		
8 Is safety equipment in working order and being used?		
9 Has any rubbish been removed from the forest?		

Auditing of forest operations should be closely connected to other processes of supervision of operations, or post-operation assessments. For example, government forest rangers may be on-site with logging contractors, and be required to complete daily monitoring records to verify progress and compliance with contractual requirements. Alternatively, forest managers may visit active compartments on an intermittent basis, perhaps weekly, to check operations. Auditing activities should aim to combine with some of these routine checks, not duplicate them.

Checklists

An important tool for auditing is a checklist. Checklists should be developed for each area of work to be audited. These can initially be very simple: a list of issues to be assessed, with results recorded as a simple yes or no answer (see Box 17.1). As experience of auditing increases, the checklists might be elaborated to provide more precise information, such as scores for performance.

Audit helps the forest manager to follow and quantify trends in operations. The more precise and sophisticated audit checklists are, the easier it will be to pinpoint problems and strengths in operations. However, the benefits of an increase in accuracy must always be considered in relation to the extra time and cost involved.

17.2 Strategic Monitoring

WHY IS STRATEGIC MONITORING IMPORTANT?

Strategic monitoring involves longer-term, repeated observations or measurements of the environmental or social effects of forest management. Strategic monitoring aims to provide information about the results of management, which can be fed back into planning and management systems. This is especially relevant to larger-scale, long-term and potentially medium to high-impact forest organizations.

Deciding which parameters should be monitored is critical. It is best to focus on a few key measurements, such as the most significant indicators of SFM, rather than accumulate and compile new data which are too voluminous to analyse or use. Monitoring has little value if the data are not analysed and fed into the forest management system.

Medium and larger-sized organizations should consider initiating a monitoring programme with detailed, longer-term monitoring. This could include:

- site productivity and stand development (growth and yield);
- water quality and quantity;
- soil erosion, compaction and fertility;
- wildlife populations, biodiversity conservation and HCVs;
- ecological interactions and regeneration of the forest;
- stakeholder activities and dependencies.

WHAT IS REQUIRED?

The information produced by monitoring at this level is often useful for forestry and biological research groups, universities and schools. Advice on appropriate techniques may most easily be found at the local university or research institute. Monitoring can provide an opportunity for an organization to collaborate with local or international researchers. For example, the forest organization could provide logistical support or accommodation for students and researchers to carry out monitoring and research. Such collaborative relationships are often a practical means for a large forestry company to provide inputs and assistance to local development in the form of opportunities for education and training.

Drawing up documented monitoring procedures may be a useful part of the development of a systematic monitoring programme. These should ensure that appropriate information is collected and analysed in a systematic and consistent manner. The following questions may be important:

- *What information is required, and why?* The forest organization's policy, objectives and targets, and environmental management programme, should be reviewed, together with any specific requirements of certification programmes. For example, one critical indicator of SFM is the effect of harvesting on growth of the residual stand and regeneration.
- *What parameters need to be measured?* Using the above example, tree diameter at breast height (dbh) and seedling counts are parameters that indicate tree growth and regeneration.
- *What is the required accuracy of information for each parameter?* For example, tree dbh to nearest centimetre and seedling count to nearest five.
- *What are the most appropriate assessment procedures for each parameter, including techniques, frequency and recording methods?* For example, measure tree dbh with a tape-measure, at a permanently marked point, every year, recorded on paper form; seedlings counted visually, within 20 x 1m^2 quadrants per hectare, six months after harvest, recorded on paper form.
- *Do any of the parameters have control limits beyond which corrective action must be implemented?* For example, if seedling densities fall below 30 per m^2, supplementary planting of seedlings may be considered.

- *How will monitoring information be analysed and then fed back into the forest management planning process?* For example, tree dbh data entered on to computer database, parameters to be calculated, statistical tests to be applied to determine growth rates.

Key areas for strategic monitoring include:

- growth and yield monitoring;
- environmental monitoring;
- social monitoring.

Growth and Yield Monitoring

Estimates of growth and yield are made based on measurements of permanent sample plots (PSPs). PSPs are a type of dynamic forest inventory and are a means of measuring tree growth, mortality and regeneration. Measurements, repeated at intervals, of the same PSPs provide information about the rate at which trees grow over time and the changes which are occurring in species composition. This information, in combination with data from static inventory (see Section 15.4) and studies of tree volumes, can be used to draw up growth and yield models which are essential for the determination of accurate sustained yield levels.[1]

PSPs can be used to monitor the effects of existing forest management practices or to measure effects of experimental treatments. A rough rule of thumb is that one plot per 1000 ha of forest is adequate.

One recommended design for PSPs suggests a square plot measuring 100 x 100 m (ie 1 ha) and subdivided into 25 subplots of 20 x 20 m. The subplots are used to measure trees, poles, saplings and seedlings: for example, all trees over 20 cm dbh might be measured in the whole 1 ha PSP. Poles (5–20 cm dbh) could be measured in a 0.2 ha subplot (ie in 5 of the 20 x 20 m subplots); saplings (1.5 m high to 5 cm dbh) could be measured in one subsample, and seedlings (<1.5 m high) in smaller subplots within that (see Figure 17.1; also Alder and Synnott, 1992).

---

1 D Alder and T Synnott, *Permanent Sample Plot Techniques for Mixed Tropical Forest*, Tropical Forestry Paper, No 25. Oxford Forestry Institute, 1992

Environmental Monitoring

Setting clear objectives and targets is important in deciding which aspects of the forest environment to monitor. The main risks identified in the ESIA should be a focus of monitoring, as should any particular environmental services that are the object of the forest organization's policy or management plan. Impacts of management on any HCVs identified in the FMU must be monitored at least annually.

Issues which environmental monitoring might cover include:

* levels and effects of soil erosion and soil fertility changes;
* impacts of forest management on water quality and quantity;
* wildlife populations, rare and endangered species and impacts on HCVs;
* effects of exotic species on the local environment.

Some aspects of environmental monitoring may need specialist inputs, at least at the design phase. Universities, national and regional research centres and institutes should be able to provide guidance or advice on what should be monitored and available techniques. Forest organizations can often benefit from cost-effective research by providing logistical support and field sites for students and researchers. Local people may be an important source of information and may have unrivalled skills in identification and surveys.

Social Monitoring

The continued assessment of local interests, perceptions and attitudes is a critical part of the monitoring programme. Again the ESIA, and the social components of the forest organization's policy and management plan, will point to the factors which will need monitoring.

Monitoring of social impacts is covered in more detail in Chapter 21.

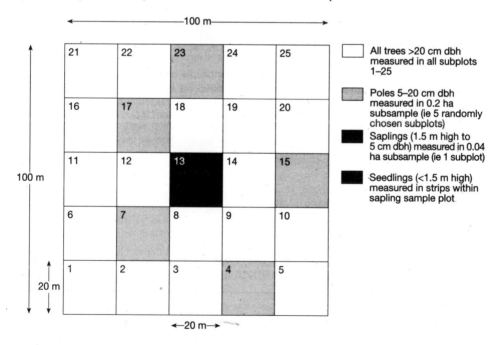

FIGURE 17.1 **A Common PSP Design**

PUBLIC SUMMARY

In order to allow informed discussion of forest management, it is good practice to make a summary of monitoring results available to the public. The FSC Principles and Criteria require a public summary of monitoring results to be made available.

A public summary should cover, at a minimum:

- yield of all forest products harvested, including timber and NTFPs;
- growth rates, regeneration and condition of the forest;
- composition and observed changes in the flora and fauna;
- environmental and social impacts of harvesting and other operations;
- productivity and efficiency of forest management.
- impacts of management on High Conservation Values.

Box 17.2 Addressing common problems with monitoring

Where do you start with monitoring? There are so many things that could be monitored and the techniques are complex and costly.

Consider trying the following:

1 Keep monitoring programmes focused on the effects of forest management: decide what you need to know in order to have a fair idea of the effects of forest management. Don't be side-tracked by esoteric or academic research!

2 Don't be afraid to use simple, cheap monitoring methods if they can provide almost the same results as more complex monitoring. For example, for wildlife surveys of specific species, well trained, organized and consistently recorded informal observations by field staff may provide information almost as accurate as a formal survey.

3 Use the resources which are available: find out whether people within the forest organization already have skills that can be used or would be interested and enthusiastic to help organize and carry out monitoring.

4 Look for easier, indirect ways to monitor. For example in a population assessment, the nests of orang-utans were counted rather than the animals themselves, because they are easier to see.

Part Five

Tackling Social Issues

Introduction to Law Five

Introduction to Part Five

Trees have always been important to people, yet social issues are often poorly understood by forest managers. Forests frequently supply a combination of goods and services to a range of different people, often at the same time. The beneficiaries of forest goods and services may be local, national or global. Sustainable forest management recognizes all the goods (for example, timber, food, fuel and medicinal plants) and services (protection of water supply, soils and even climate) which people obtain from forests, including:

- wood products;
- non-timber forest products (NTFPs);
- watershed conservation;
- soil conservation;
- wind and noise control;
- natural scenery;
- recreation and eco-tourism;
- culture and religion;
- microclimate regulation;
- climatic stability;
- carbon storage;
- maintenance of biological diversity and High Conservation Values in forest ecosystems.

The word 'stakeholders' has become used in recent years to describe all the people who are (directly or indirectly) affected by, or interested in, forest management. It is because so many people are stakeholders, and have an interest in the way forests are managed, that social issues are often the biggest challenge for forest managers. Many markets and authorities are calling for 'people-friendly'

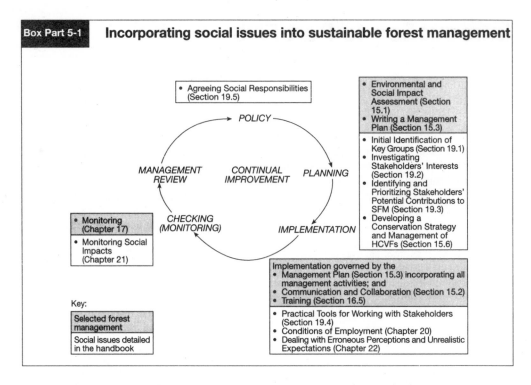

Box Part 5-1 **Incorporating social issues into sustainable forest management**

forest production. Part Five looks at commonly occurring social challenges, and suggests some ways of working with stakeholders.

Social concerns need to be incorporated at all stages: planning, implementation and monitoring. Figure Part 5-1 shows the ways that the social issues discussed in Part Five can be incorporated into forest management activities.

Most of the techniques explained in Part Five have already been used successfully in community forestry projects, and in agriculture – though fewer of them have been tested in commercial forestry. Because consideration of social issues is still relatively new, there are few well-documented experiences of good management of social impacts in commercial forestry. However, the approaches outlined in Part Five represent some of the best ways of making progress, based on current understanding. Part Five includes information on:

Chapter 18 Why Social Issues are Important

Chapter 19 Working with Stakeholders

Chapter 20 Conditions of Employment

Chapter 21 Monitoring Social Impacts

Chapter 22 Dealing with Erroneous Perceptions and Unrealistic Expectations

18 Why Social Issues are Important

In the past, forest management training emphasized technical and economic approaches, but nowadays foresters are also being asked to focus on social issues, as part of a sustainable forest management process. For example, the FSC has developed a social strategy with a set of core social values (see Box 18.1). Resolving social issues requires an understanding of the values people hold, and the participation of key groups of people in making choices between forest goods and services.

Social issues can be contentious, particularly if there is involvement by different interest groups in forestry, resource rights, and broader social concerns, such as those affecting people beyond the forest area.

Part Five provides more information about the issues summarized in Section 15.2 and will help to bring forest managers up to date with ways to approach social issues and achieve successful collaboration. To stay up to date, try developing links with some of the organizations shown in Appendix 4.1.

Forest organizations are recommended to make a special effort to address social issues because of:

Box 18.1 **Social strategy of the Forest Stewardship Council**

The FSC states in its Social Strategy[1] that its 'core social values' are:

- **Access:** Strive to make certification (see Part Six) equally accessible to all forest owners and managers, regardless of age, gender, ethnicity, faith, cultural background, geographical location, scale or intensity of the operation, or ecosystem in which they operate.

- **Partnerships:** Build long-term partnerships based on transparency, respect, mutual learning and reciprocal accountability.

- **Legal rights:** Support and facilitate the legal and meaningful recognition of indigenous peoples', local communities' and workers' rights, including traditional and customary rights to forest-related knowledge.

- **Equity:** Promote inter-generational equity and the equitable distribution of benefits from the forest to Indigenous peoples and local forest-dependent communities

- **Cultural identity:** Respect cultural identity and diversity, traditional local governance structures and decision-making processes, and the right to self-determination and self-development.

- **Subsistence forest use:** Recognize and support as fundamental the subsistence use of forests by forest dwellers.

- **Traditional forest stewardship:** Support traditional forms of good forest stewardship and their adaptation to changing social, economic and environmental conditions.

---

1 Forest Stewardship Council, FSC *Social Strategy: Building and Implementing a Social Agenda*. Version 2.1. Forest Stewardship Council, Bonn, 2003. www.fsc.org

- people's rights to forests;
- good business practice;
- ethical reasons;
- legal reasons;
- responding to external pressures;
- contribution to sustainable development.

18.1 People's Rights to Forests

People may possess a range of rights which directly concern the forest and its use. These rights include:

- territorial rights;
- ownership of trees and other resources, such as minerals, grazing and wildlife;
- rights of access, use and control (for specific purposes and times);
- intellectual property rights.

However, this range of rights can generate complexities for forest managers; for example, if many different rights overlap on one territory. In addition, many rights may not be established in legal documents, especially traditional rights.

Other relevant stakeholder rights relate to *community cohesion*. For example:

- rights to protection of cultural heritage, landscapes, and folklore;
- collective rights, for example a community's rights to self-determination and to represent itself through its own institutions (see Box 18.2);
- religious freedom;
- rights to development;
- rights to privacy.

Box 18.2 **Rights of local communities**

The following rights may be assumed for all local communities with legal or customary tenure over land areas with forest resources:[2]

- the right to participate in defining how the local community can control management on their lands;

- the right to free and informed consent, including the right to grant, withhold or withdraw consent;

- the right to delegate consent, including the right to set conditions of delegation (eg to set performance benchmarks) and the right to revoke delegation;

- the right to protect their rights;

- the right to self-define their own community and their own forest resource needs.

---

2 R Collier, *Development of Draft Guidance on the Interpretation of* FSC *Principles 2 and 3*. Forest Stewardship Council, Bonn, 2004. www.fsc.org

Some groups are forceful in exerting their rights, or have the skills and resources to do so. Others are less well equipped. Irrespective of the stakeholders' powers and resources, a forest organization is likely to experience problems if it ignores or violates stakeholders' rights.

18.2 Good Business Practice

When social issues are not addressed, different combinations of conflicting interests, unfulfilled demands and inadequate responses, have led to problems for forest managers, such as:

- legal challenges and compensation claims;
- unrealistic demands for infrastructure development;
- uncooperative local and political leaders;
- withdrawal of labour and confrontation with management;
- stalled negotiations with neighbouring communities, politicians and employees;
- claims on land after it has been developed for forestry;
- bad press and political backlashes;
- loss of markets and contracts;
- vandalism and sabotage;
- hostility towards forest managers and companies.

In contrast, people contribute positively, or take responsibility, for forest management when there is a benefit – or incentive – to do so. This usually means collaborating to identify these incentives (see also Section 15.2).

Working to accommodate and balance the legitimate interests of different groups is good business practice. By taking an active approach to people as well as trees, forest managers greatly increase the potential social benefits from forest management, and their chances of support from others. They also increase their ability to anticipate future developments, for example in legislation or the market, and are more likely to gain competitive advantage from the situation. An active approach requires skills, commitment and discipline for it to work.

18.3 Ethical Reasons

People living in forests, or the agricultural areas around them, are often poor and may be vulnerable to the impacts of more powerful groups. They may be indigenous peoples, or those with long historical dependence on the forest, or landless people who have migrated from other areas. Some forest managers may seek to exclude local people's access to the forest. Yet more equitable distribution of opportunity and wealth is a goal of many countries and societies. Better forest management all round can result where local people have secure access agreed with the forest manager.

There are few more contentious issues than the use of land. Land use is at the root of many cultural and political beliefs. Commercial forestry, with its requirement for the use of large areas of land for long periods of time, can set off this political 'minefield'. In most countries, the granting of rights to utilize forests for commercial purposes is accompanied by the acceptance of social responsibilities for contributing to the wellbeing of people and reducing inequality. Some of these responsibilities will be written in concession agreements and contracts by governments who set the terms. Others may only become clear through collaboration with local stakeholders, such as respect for cultural sites.

18.4 Legal Reasons

Many interests in forest land are expressed as legal rights, for example in national forestry legislation. However, local people's own use of trees almost always comes last in any list of priorities expressed through local and national laws, or plans for forest lands. This problem may be made worse by laws affecting resources on, or under, forest land, such as minerals (mining laws), soil (land taxes) and access (public rights of way). National laws may also cover other social issues linked to forests.

Internationally, there is a growing body of legally binding governmental agreements which states may be party to. These include:

- International Labour Organization (ILO) Convention on the Right to Organize and Collective Bargaining (1949);
- ILO Convention on Occupational Safety and Health (1981);
- UN Covenant on Economic, Social and Cultural Rights (1966);
- UN Covenant on Civil and Political Rights (1966);
- UN Declaration on the Human Right to Development (1986);
- ILO Convention 169 concerning Indigenous and Tribal Peoples in Independent Countries (1989);
- UN Convention on Biological Diversity (1992).

Laws on land, environment and forestry are being reviewed in many countries, especially the links and conflicts between formal and customary rights. Significant efforts are also under way in many countries to improve the effectiveness of the law in preventing the negative impacts of bad forestry, and also to avoid the perverse effects of law and its enforcement on local livelihoods and the forest. Many legal initiatives are based on new legal principles which come from international consensus on sustainable development (such as inter-generational equity and the polluter-pays principle). Anticipating these types of developments, by working proactively with governments and others such as relevant university departments or research institutes, may bring competitive advantages to forest organizations.

18.5 Responding to External Pressures

External factors may apply pressure, or place responsibility, on the forest manager to address certain social needs and expectations:

- *Pressure from forest people's groups and their initiatives.* Demands may include greater respect for local populations' rights and support for various projects.
- *Pressure from local groups to contribute to social and economic development.* Demands may include support for local organizations, employment opportunities, and infrastructure.
- *Pressure from trade unions and their initiatives.* Demands may include greater respect for workers' rights to organize and negotiate, fair wages and benefits, health and safety at work, and the right to skills development throughout the organization.
- *Responsibility to ensure forest people's appropriate rights to monitor, control and negotiate.* Demands may include the need to develop agreements, manage conflicts and compensate for losses.

18.6 Contribution to Sustainable Development

Sustainable development is concerned with economic, environmental and social goals, and the links between them. Forest management which addresses social issues, as well as implementing good silvicultural techniques, can play an important role in development. However, a forest organization alone cannot be expected to achieve those fundamental social goals, such as equitable development, which are (or should be) set at national level. For this reason it is useful to clarify the needs and wants of stakeholders at an early stage, in order to establish if any group has unrealistic expectations about the forest operation (see Chapter 22).

19 Working with Stakeholders

Demands on forests are made by many different groups of people and organizations. Stakeholders' actions can have positive and/or negative impacts on forests, forest management and other stakeholders. In addition stakeholders' lives can be affected positively and/or negatively by forest operations.

There are some stakeholder groups upon which the success of the forest organization will depend. Forest managers need to endeavour to satisfy such stakeholder groups. In practice this means understanding and working effectively with these groups and minimizing, or dealing with, conflicts. Some stakeholders relate to, and are more involved in, certain objectives of forest management than other stakeholders.

Once all stakeholders are identified, a process is needed for establishing two key groups:

1 stakeholders who have useful contributions to make to the success of the forest organization;
2 stakeholders who will be most affected by forest management.

These two groups may overlap significantly since some of the people who are most affected by forest management may also be those with the most to offer. Forest managers should aim to work most closely with these key groups.

The forest organization needs to be able to interact, and build up trust, with the different stakeholder groups. Skills in this area are new and evolving. A step-by-step approach, based on participation, will help in building up these relationships.

Effective participation takes time. For it to work, stakeholders need to be identified, communication between the forest organization and stakeholders must be established and gradually relationships can be formed and improved.

Tools for carrying out these key tasks are covered in the following sections:

19.1 Initial identification of key groups
19.2 Investigating stakeholders' interests
19.3 Identifying and prioritizing stakeholders' potential contributions to SFM
19.4 Prioritizing key stakeholders
19.5 Practical tools for working with stakeholders
19.5 Agreeing social responsibilities
19.6 Company–community forestry partnerships

19.1 Initial Identification of Key Groups

Stakeholders are defined as any people who are directly or indirectly affected by or interested in forest management. This includes a wide variety of people: some examples are shown in Figure 15.3.

There are many ways of starting to identify stakeholders (see Box 19.1). What works in one situation may not be appropriate in another, and the degree of detail needed will vary depending on the numbers and interests of stakeholders.

Box 19.1 **Approaches for identifying stakeholders**

Each of the approaches listed below has advantages and risks. To overcome the risks of missing key stakeholders, the forest manager needs to consciously work to involve all stakeholders. Using a combination of approaches will reduce the risks associated with any one particular approach.

Self-selection. Staff of the forest organization make announcements at meetings and/or in newspapers, local radio or use other local means of spreading information. Groups and individuals then come forward and ask to be involved. The approach works best for people who already have good contacts and see it in their interest to communicate. Those who are in more remote areas, or are poor and less well educated, and those who may be hostile to the forest organization, will not come forward in this way.

Self-selection

- *There is a risk that local elites, or some with inequitable objectives, put themselves forward.*

Identification by forest organization staff. Those who have worked in the area for some time can identify groups and individuals who they know to have interests in forest issues and are well informed about them.

- *There is a risk that the same people will always be selected. It may also lead to the under-representation of women.*

Identification by forest organization staff

Identification by other knowledgeable individuals. To make contact with individuals and groups who have had little contact with the forest organization, or who have been hostile to it, knowledgeable advisers from other organizations and communities can help. For example, land and agricultural agencies may be able to recommend relevant farmers and settlers; local government, religious and traditional authorities; forest agencies and other forest organizations may all be able to identify key representatives of different forest interests.

Identification by other knowledgeable staff

- *There is a risk that less vocal groups will be under-represented.*

Identification through written records, and population data. Forest organizations often have useful records on employment, conflicting land claims, complaints of various kinds, people who have attended meetings, and financial transactions. Forestry officials may have important historical information on forest users and records of permit holders. Census and population data may provide useful information about numbers and locations of people by age, gender, religion, etc. Contacts with NGOs and academics may reveal relevant reports and information about knowledgeable or well-connected people.

Identification through written records

- *There is a risk of putting too much emphasis on inaccurate and incomplete data, and of using information that is oversensitive.*

Identification and verification by other stakeholders. Early discussions with the first identified stakeholders will reveal their views on the other key stakeholders who matter to them. This will help the forest organization to better understand stakeholder interests and will enable identification: firstly, of key individuals in the groups who act as representatives; and secondly, of other groups who need to be contacted or considered.

Identification and verification by other stakeholders

- *There is a risk that a small group of stakeholders will nominate each other, to the exclusion of others.*

Some of the key questions to ask when trying to identify stakeholders are:

- Who is or who might be affected, positively or negatively, by the forestry organization?
- Who are the representatives of those who are/might be affected?
- Who is likely to resent any aspect of forest management and mobilize resistance against it?
- Who can make forest management more effective through their participation or less effective by their non-participation or opposition?
- Who can contribute resources and information?
- Whose behaviour has to change for forest management to succeed?

At an early stage it should be possible to make a first attempt at identifying stakeholders. Often there will be two distinct groups:

1 *Those who have rights in the forest or forest organization.* Those who perceive themselves to have rights are likely to include stakeholders who are dependent on the forest or the forest land for their survival. They may have few other ways of making their livelihood, so they may be strongly affected by forestry operations. The concerns of stakeholders with such rights must be addressed.
2 *Those who have interests in the forest or forest organization, but not necessarily any rights.* This includes all other people who have a stake or interest in the forest or area. Various people may have the means to influence forestry operations either positively or negatively. The concerns of stakeholders with 'interests' may, or may not, need to be addressed. This is discussed further below (see Section 19.3).

This is the appropriate point to start developing more information through Environmental and Social Impact Assessment (see Section 15.1) and through initiating the process of communication and collaboration (see Section 15.2).

Scale considerations! ♀
Sometimes it is possible to identify who has rights and who has interests in the forest simply through discussion with stakeholders. This might include contact with local authorities (both formal and traditional) and local representatives of government forestry and agriculture agencies, as well as making informal contacts with people from local villages.

19.2 Investigating Stakeholder Interests

Stakeholders have different interests, and investigation of these through discussion can help identify how people view their current and potential roles in forest management. Some examples of different stakeholders and their interests in the forest are shown below in Box 19.2.

Finding out how people see their own roles in forest management is an essential step towards agreeing the objectives of SFM. One way of doing this, which has been found to be useful, is to focus discussion on stakeholders' *rights, responsibilities and returns* ('*three Rs*') with respect to particular activities in the forest (for example, hunting, farming, infrastructure development and transportation).[2] An example of the use of this approach, in developing a joint agreement about hunting, is shown in Box 19.3.

As a result of their differing rights, responsibilities and returns, stakeholders also have different sorts of *relationships* ('*the fourth* R') with each other. Some may not be aware of each other, or may ignore each other; others may be in varying states of *disagreement or cooperation* over different issues related to forest management. An open communication process to clarify existing disagreements

---

2 O Dubois, *Capacity to Manage Role Changes in Forestry: Introducing the '4Rs' Framework.* IIED, London, 1998

and cooperation levels between stakeholders can help forest managers develop an understanding of the types of action necessary to build useful cooperation, to deal with problems while they are still small, and to find solutions which satisfy all stakeholders.

For example, if it is known that a local group has had its access to certain timber trees removed by a forest organization, while another group's rights to take certain trees has not changed, the company might need to defuse the resulting disagreement, and establish useful cooperation. One solution could be to plan to build an access road in a way which benefits the affected group.

Some of the key questions to ask when investigating stakeholder interests might include:

- What are the stakeholders' experiences and expectations of the forestry organization?
- What benefits or costs have there been, or are there likely to be, for the stakeholder?
- What stakeholder interests conflict with the goals of the forestry organization?
- What resources has the stakeholder mobilized, or is willing to mobilize?

Disagreements between stakeholders may occur between groups which seem similar, for example different communities living in the forest or different groups of environmentalists. Figure 19.1 shows the way in which a matrix of stakeholder groups can be used to show possible disagreements between different groups, in this example over harvesting trees from a particular area of forest. Dark cells indicate a major disagreement, lighter cells indicate a minor disagreement, and clear cells indicate little or no perceived disagreement.

Broad groupings of stakeholders, for example 'people who live in the forest', in practice usually consist of smaller, and less cohesive, groups with different interests, rules and leaders. Some people will use timber trees, others may be more interested in hunting animals. Women's interests may also be different to men's. This is highlighted by the matrix in Figure 19.1 in which some people who live in the forest are in disagreement with other people living in the forest. Identification of stakeholders which stops at the level of the larger groupings will not be very useful. Broad stakeholder groups therefore need to be 'unpacked' into smaller groups of people who have major interests or rights in common.

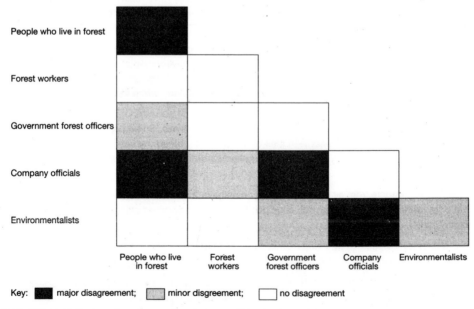

Key: ■ major disagreement; ▨ minor disgreement; ☐ no disagreement

FIGURE 19.1 Example of a matrix of possible stakeholder disagreements over tree harvesting

Box 19.2 **Some examples of forest stakeholders and their interests**

People who live in or near the forest. For example, ethnic groups which have lived in the area for many years. They may have useful knowledge about their environment and indigenous forest management practices. However, they often have less power and money than other groups, and tend to be the main losers if commercial approaches to forest management are implemented without thought for their livelihoods.

People who live further away and who come to the forest in times of stress, or at certain seasons, for example, to gather forest products, graze livestock and so on. They may have an opportunistic outlook and varying levels of indigenous knowledge.

Settlers from elsewhere in the country, or from other countries. Typically poor and often unfamiliar with ways to thrive in forest environments, especially if they are more familiar with agricultural lifestyles. They may practise slash-and-burn farming and be expecting to move on to new areas.

Forest workers are typically thought of as men seeking wages and other direct benefits. However, women also work in both the formal (eg nursery workers) or informal (eg domestic firewood collection) sectors. Forest workers may be far from their families and other traditional sources of social support or control.

Small-scale entrepreneurs have information and money which enable them to engage in the marketing and processing of forest products. They may also run small-scale logging operations or be contracted to do forestry-related work for larger companies.

Forestry officials work for the government, often in bureaucratic systems with much paperwork. They may look down on other groups as being 'in the way' of their mandate to protect and administer the forest. Some may also use their position for extra personal gains.

Managers of forest organizations. Often from urban areas and may have much power over other groups in the forest area. They may possess good forest knowledge for timber production; but may have poor knowledge of other forest goods and services and local social values.

Environmentalists. Usually urban-based, concerned with putting pressure on other groups, particularly the government and timber companies, to improve environmental sustainability. May be influential through influencing votes, letter-writing, demonstrations, fund-raising, obtaining media coverage, organizing local resistance and spreading information worldwide through their links with international organizations.

Politicians participate in decision-making at local and national levels. Some politicians may be influenced by the preferences of people in their constituencies; others by national level aims which may clash with local needs. They may also be targeted by both environmentalists and the timber industry.

National citizens. People from all over the country may have a voice in forest management – by voting or by writing letters they can influence politicians and government officials; their interests may be highly varied.

Global citizens represented through inter-governmental agreements and markets; for example, international agreements on climate change may affect national forest management.

Consumers who buy forest products. They are usually most concerned about affordable supplies of products, but some want proof of sustainable forest management practices.

Box 19.3	**'The Three Rs': an example of stakeholders' rights, responsibilities and returns for hunting**

The 'three Rs' can be used as the focus of discussion in a public stakeholder meeting or in one-to-one meetings with each stakeholder group. It works best if local livelihood topics, such as hunting or farming, or forest management operations, such as road building or monitoring, are dealt with separately.

The following example shows a 'three Rs' matrix as the outcome of a discussion of a situation where only hunters originating from the villages neighbouring a forest management unit are allowed to hunt certain wild animals.

Stakeholders	Rights	Responsibilities	Returns
Village hunters	To hunt certain amounts of certain species in the forest	To avoid over-hunting and protected species	Money from sale of game meat
	To use the company's transport in cases of emergency	To help identify and prevent illegal logging, hunting, etc	
Company officials	To monitor hunter activity	To guarantee respect of the government's regulations	Better nutrition of workers with meat
	To help prevent access of hunters without rights from other villages		Better relations with villages
Forest workers	Priority or discount in buying game meat	To prevent hunters' use of company's transport, except in emergency	Cheaper and regular supply of game meat
		To help communication between hunters and company	
Government forest officers	To check monitoring of hunters	To ensure other stakeholders are aware of government regulations and to ensure compliance	Controlled levels of hunting by villagers and prevention of illegal activities
	To prosecute illegal hunters To use the company's transport in cases of emergency		

19.3 Identifying and Prioritizing Stakeholders' Potential Contributions to Sustainable Forest Management

As well as having different interests, stakeholder groups have differerent abilities and potential contributions to various aspects of SFM. Some stakeholders might be able to make large contributions in some activities, but have no interest or ability to contribute in others.

The matrix shown in Table 19.1 is an example of a method for guiding the identification of key stakeholders with potential to contribute to aspects of SFM. The matrix assigns scores to stakeholders on their potential contribution. The average score for each stakeholder is then calculated. A cut-off point can be decided upon to help indicate who are the key stakeholders. In this example, if the cut-off point is 'greater than 2', then farmers near the forest and the environmental NGOs would not be regarded as key stakeholders.

This method does not provide cut-and-dried answers about who the key stakeholders are, but it may assist the forest manager to decide which stakeholders have the most potential to contribute to discussions on particular forest management activities. The manager needs to keep in mind, however, that this method does not consider which stakeholders should be considered for reasons other than their potential to contribute to forest management: other criteria such as potential impacts of the forest management activity on the stakeholder group, legal, ethical or political reasons may also make it necessary to involve a particular stakeholder group in discussions.

Some of the key questions for assessing potential contributions to SFM include:

- Who is dependent on whom?
- Which stakeholders are organized?
- Who has control over resources and information?
- Which problems, affecting which stakeholders, are to be prioritized?
- Which stakeholders' needs, interests and expectations should be given priority attention?

This brings us into the tricky area of the different degrees of power and influence which stakeholders possess (see Box 19.4). For example, those who live in, or near, the forest often have little power compared to other stakeholders whose education, wealth or job enables them to have a greater say in forest management. This may have negative effects on local livelihoods, and also means that the useful knowledge and abilities of these local people are not contributing as they could to improving forest management. These power differences need to be recognized, and a process of good communication can lead to agreement that different stakeholders will be treated in different ways.

TABLE 19.1 Example of a matrix for identifying potential contributions to SFM

Forest Management Activity	Stakeholder Group				
	Indigenous forest dwellers	Farmers near forest	Local traders in forest products	Forestry workers	Environmental NGOs
Management planning	3	2	3	2	1
Inventory	3	1	3	3	2
Road building	1	3	2	3	1
Harvesting	2	2	2	3	1
Conserving biodiversity	3	1	3	2	2
Monitoring	3	2	3	3	2
Securing tenure	3	2	2	2	1
Protecting culture	3	2	2	2	2
Conditions of employment	2	2	2	3	1
Local development	2	3	3	2	2
AVERAGE SCORE	2.5	2	2.5	2.5	1.5

Key: 3 = High; 2 = Medium; 1 = Low potential contributions to forest management

19.4 Practical Tools for Working with Stakeholders

Once the key stakeholders have been identified, the forest organization needs to establish a process or framework for communication and collaboration with them. As a first step, the forest management team may consider developing a social policy to set social objectives and to give scope for continual improvement, input to planning and feedback on operations.

A social policy may be developed as part of the sustainable forest management policy described in Chapter 10. It should set out overall social objectives of the forest organization and provide the basis for setting objectives and targets (see Section 19.5). A social policy must be drawn up in collaboration with the identified key stakeholders. At present, few forest organizations – except some of the largest – have a formal social policy, although many will have evolved an informal or partial approach to stakeholders.

In drawing up the social policy, useful guidelines to bear in mind are:

- Establish useful and *regular two-way communication*. This can be coordinated through a forum or liaison group, if the operation is large and there are many stakeholders.
- Develop an *understanding and appreciation* of other stakeholders' resource management and decision-making traditions.
- Encourage *sharing of skills and technology*.
- Value *transparency and trust* among stakeholders, and try to build it.
- Employ *staff who understand, and are respected by*, other stakeholders – which will mean some who themselves come from these groups.
- *Stay up to date* on local legislation and best practice concerning employment, health and safety.
- *Make contact with best-practice companies*, and participatory forest management projects, to share lessons.
- *Open up the forest to visits* from stakeholders, and to constructive discussions.
- *Consult stakeholders in detail* when contemplating afforestation, infrastructure development, and major changes to forest management and employment practice.
- *Avoid rushing*. Be realistic about the time it takes for stakeholders to consult among their own constituencies.

Specialist assistance may be needed by larger forest organizations which have many stakeholders. In some cases this may need to be hired in for identifying, planning and managing appropriate processes and responses.

New skills and attitudes may be needed by the forest management team for working with stakeholders. Training may be needed in areas such as:

- cultural and gender awareness;
- chairing of meetings and workshops;
- facilitation, for example of processes involving several stakeholders;
- consensus building and conflict management;
- how to ask open questions and develop discussion;
- developing enthusiasm and commitment;
- knowing when and how to bring in third parties.

Many of the participatory appraisal principles and methodologies which specialists may use, such as constructing community profiles, mapping and priority ranking exercises, are useful for developing participation and communication with different stakeholder groups. These methods are listed in Box 19.5.

| Box 19.4 | Tricky issues – power differences |

It is not easy for forest managers to sit down with other stakeholders and talk about power differences between them! However, it should be recognized that stakeholders regularly weigh up each other's power in their day-to-day communication, often without recognizing it. This is reflected in the way people communicate, and the issues they feel comfortable communicating about. Sometimes it is necessary to take a deeper look at what is going on, for example because:

- some stakeholders are losing out from the current arrangements;

- potential positive contributions to forest management and potential actions which may undermine it may be hidden under the current arrangements;

- important issues are just not being talked about;

- there are problems which might be solved if stakeholders were free to look at them in another way.

So, as well as identifying stakeholder interests and the impacts they have, it may be necessary to get a feel for the processes by which stakeholders pursue their interests. This involves looking at the sources of power, the means to pursue interests and the range or degree of power which stakeholders have:

- *Sources of power* might include: having useful personal contacts; earning money; possessing scientific knowledge; holding an important job; owning land; controlling equipment or vehicles; having authority to provide loans, allocate budgets or hire and fire employees; securing international or political support.

- *Means to pursue interests* might include: legal means, eg the rights to farm, or to extract timber; illegal means, eg bribery or sabotage; formal means, eg regulations or public meetings; informal means, eg forming alliances or lobbying.

- *Range or degree of power.* The extent of influence may be very local, affecting just a few people or a small area of forest, or it may be far-ranging, affecting many people over a large area or the whole country.

Frameworks such as the 'four Rs', described in Section 19.2, and stakeholder power analysis,[3] can be very useful in indirectly raising power issues in ways which are acceptable. For example, a discussion on how forest workers depend on the company for their jobs and income may reveal how dependent the company is on the workers for carrying out forest operations, and on villagers and workers for reducing the risk of trouble or dispute. All depend on government forest officers to explain and enforce regulations, but the officers are often highly dependent on the company for the means to do this. Recording the main themes and issues in such indirect discussions of power will be likely to prove useful in the course of forest management.

Recognizing power differences does not always mean that forest managers and other stakeholders can or should do anything to change them. But to be open about the differences is important. Some differences can be acknowledged in joint agreements (see Section 19.5), while identifying problems stemming from power differences may give stakeholders ideas on how to improve their knowledge or capacity to address the problems. Power differences usually change over time.

---

3 J Mayers, *Stakeholder Power Analysis*. Power Tools Series No 2. IIED, London, 2001. Available at www.iied.org/forestry/tools

Box 19.5 Some social tools for working with stakeholders

Brainstorming. Generate a list of issues in a stakeholder group where 'anything goes'. All ideas must be welcomed, from anybody in the group. One person can record all the ideas on paper or a blackboard, or all group members write down their ideas on different pieces of paper which are then stuck on a wall. Discuss how to sort and prioritize the issues.

Interviewing involves questioning and listening to people. Some questions and topics can be thought of in advance; others can arise during the interview. Use a guide or checklist (see below) to pose questions and investigate topics as they arise. Write up notes during or straight after the interview.

Checklists are informal lists of issues to investigate. Use these to guide interviews, in place of formal survey methods or questionnaires. Regularly modify them as new information and issues arise.

Village/community meetings. Attend existing village or community groups if they are broadly representative; or call special meetings to give out information and to get feedback. Communicate intentions of a forest organization at such meetings – especially in the early stages of identifying stakeholder groups and possible impacts. Such meetings are essential when community-wide issues or conflicts emerge.

Focus groups. Convene special groups to discuss a particular topic. For example, farmers wanting land within the forest or hunters and their practices.

Key informants. Take time to consult in depth with acknowledged local experts, elders or knowledgeable individuals who can provide critical information.

Direct observation. Look at situations for yourself and ask questions directly related to what you see.

Participatory mapping. Provide opportunities for stakeholders to prepare maps of resources/problems/conflicts. This can be done on paper or blackboards, or can use local materials such as sticks, leaves, stones, grass, coloured sand, cigarette packets, etc, on the ground. Allow one map to lead to others, as more and more people get involved. Encourage interruption of map preparation to enable more focused discussions to take place. A range of maps can be produced, such as:

Box 19.5 **continued**

- *Resource maps* – depicting villages, forests, farms, hunting grounds and so on.
- *Tenure and rights maps* – indicating who owns, and has rights to, which areas or resources.
- *Impact and action maps* – recording where particular impacts occur or actions that are needed.
- *Mobility maps* – showing people's movements to other towns and cities from their community. These can reveal valuable information about seasonal movements, markets used, transportation difficulties and so on.

Time lines. Work with groups to prepare a history of major recollected events in a community with approximate dates, and discussion of which changes have occurred and why (cause and effect).

Matrix scoring. Use matrices to agree ordering and structuring of information, and then for planning. Agree ranking criteria (matrix rows) and relevant issues (matrix columns). Ask stakeholders, usually in a group, to fill in the boxes for each row.

Group contracts. A formal written contract in which a group's members set out their roles and responsibilities, and what they see as appropriate behaviour and attitudes towards one another and towards other groups. Ensure the contract is seen as a working agreement between all group members. This might be appropriate for outgrower schemes and the forest organization's own liaison committee/group.

19.5 Agreeing Social Responsibilities

SFM is about achieving an agreed balance between economic, environmental and social objectives. Sometimes many of these objectives can be met; at other times, choices will have to be made between them. The choice is not merely an economic or technical one. It is a choice concerning values, especially as many of these values are not covered by market decisions, such as biodiversity, equity and aesthetics. Some examples of possible positive and negative social outcomes of forest management are shown in Table 19.2.

Using some of the tools outlined in Section 19.4, participation of key stakeholders should be encouraged in setting SFM policies, objectives and targets. The process of participation and agreement aims to ensure that:

- the resulting policy, objectives and targets are credible;
- all stakeholder groups know who was involved in developing them, and have confidence in the procedures used to negotiate them;
- a broader range of ideas, skills and inputs is reflected;
- many perspectives are heard;
- important information is brought to the table, that may otherwise have been unavailable;
- a best consensus on trade-offs is achieved, considering current conditions and knowledge in the area in question;
- a stronger foundation of stakeholder trust and accountability is built;
- partnerships, processes, and commitment to implementation are strengthened.

The process of stakeholder participation in setting SFM policies, objectives and targets should result in an agreement of social responsibilities, outlined in the social policy. This should be made publicly available, in order to make clear what can and cannot be expected from the forest organization.

TABLE 19.2 **Examples of possible social outcomes of forest management**[4]

	OUTCOMES TO AIM FOR	*OUTCOMES TO AVOID*
Rights and tenure	Recognize people's rights to land, trees, other goods and services. Support the exercise of local rights.	Clear rights for some may exclude uses of forest by others. Disregard for customary rights and increased landlessness.
Living standards	Support local employment and training. Contribute to household income and livelihoods. Ensure adequate health and safety provisions.	Lack of employment rights and security. Poor groups lose access to land and forest products for food, fuel and income. Increased accident rates.
Infrastructure	Design roads to improve transport/ communication for multiple purposes. Improve housing, education and health facilities.	Roads and machinery cause accidents and pollution. Facilities disregard cultural norms.
Population	Ensure that any immigrant labour brings new skills and market opportunities.	Expulsion of indigenous population from FMU. Settlers take land and forest resources, and disrupt traditional economy – competing excessively with locals. Dependency on short-lived forestry employment. Increased prostitution and social problems.
Cultural values	Record, recognize, ensure access to, and help maintain religious and cultural sites.	Loss of, or damage to, culturally important sites. Conflicts between FMU priorities and cultural values. Reductions in aesthetic and recreational values.
Community and worker organizations	Assist establishment or strengthening of cooperative organizations. Encourage organized and effective unions. Support marginal groups eg indigenous populations and women.	Local communities' priorities not understood or supported. Small-scale producers unable to compete. Only some sections of the community favoured.
Distribution of benefits and costs	Work with local groups to ensure equitable sharing of costs and benefits.	Poor and landless do not get benefits. Conflict over employment opportunities.

Social responsibilities will vary according to the local situation. They depend on:

- the type of tenure and rights;
- the level of services already provided by the government;
- compensation of the workforce;
- the size of the forestry operation;
- local expectations.

The appropriate social responsibilities for the forest organization to take on can only be defined through the consultation and participation process described above. However, some ideas about

4 Developed from Shell/WWF *Tree Plantation Review*, Study No 7. Shell International and WWF, 1993

Box 19.6 **Ways to contribute to local livelihoods and development**

1   Provide support to those local decision-makers who can address community needs.

2   Provide support during transition to a cash economy (if appropriate).

3   Provide employment opportunities for members of local communities at all levels of the organization.

4   Provide training and skills development.

5   Ensure long-term security of employment.

6   Pay compensation for forest resources lost to the community, for example provide/subsidize building materials where access to timber trees has been stopped.

7   Purchase local goods and services, supporting local organizations.

8   Produce-sharing, eg establishing the community's rights to a certain percentage of the timber harvested, to other forest products, or to thinnings in a plantation.

9   Develop income-generating opportunities, for example outgrower schemes in which local people produce forest products for sale to the forest organization.

10 Develop profit-sharing or share-holding options in the forestry organization.

11 Provide infrastructure for local communities (for example roads, schools, hospitals, and communications) based on needs identified by the community.

---

ways in which in the forest organization may be able to contribute to local community livelihoods and development are listed in Box 19.6. Although these may provide a starting point for discussion, the forest management team needs to be aware that there may not be support from stakeholders or options available to the forest organization.

In agreeing an SFM policy with stakeholders, some key steps to cover are:

- *Identify legal and traditional resource rights and tenure* and agree rights and responsibilities for their management.
- *Identify areas of cultural, historical, economic or religious significance* and socially important High Conservation Values (see Box 15.18) to the community which may be excluded from forest management activities.
- *Identify cultural practices which should be respected,* such as taboo days, festivals, and use of forest products in cultural events.
- *Identify practical socio-economic considerations which should be respected,* such as the location of roads and extraction routes (in relation to housing and other activities), protection of drinking water sources, timing and planning of logging operations so as not to disrupt farming activities.
- *Identify existing community infra-structure and agricultural crops which should be respected,* for example making agreements to maintain existing bridges used in forest operations, or compensation for damage to crops.
- *Agree who can monitor operations and how* (see Chapter 21). This may include monitoring by communities, the forest organization, government officers or local NGOs.
- *Agree sanctions for failure to comply with agreed actions,* to ensure that all sides comply. These may be agreed and enforced between forest management and local communities or may involve external sanctions by government forest officers or others.

- *Document the process and results*, including a record of who was involved, what was agreed and responsibilities for compliance.
- *Sign an agreement with witnesses and make this public.* Witnesses may include government forest officers, local leaders and representatives of other major stakeholders in the FMU.

Scale considerations! ♀

For large organizations the process of agreeing social responsibilities will require specialist inputs (such as might be found in local or national universities or research institutes, or through some of the institutes noted in Appendix 4.1), as the issues will be complex and the consequences of misunderstandings may be severe. Large organizations, intentionally or unintentionally, create high expectations of the benefits which they will provide: these expectations need to be dealt with sensitively.

19.6 Company–community forestry partnerships

All sorts of deals have been struck between forestry companies and local communities over the years. Companies have sought to make deals to secure access to land and labour, and continuous supplies of wood – as well as to demonstrate their good-neighbourly intentions. Communities have sought employment, technology, infrastructure, social services and sources of income – and secure access to a wide range of forest products.

A range of factors may determine whether companies and communities strike up deals or actively avoid them. For companies, external policy or market duress to practise fair trade or sustainable forest management may be important, as may economic considerations, such as the potential to cut costs, share risks or gain access to resources through engagement with local groups. Companies can provide skills, technologies, resources and access to markets that the community would otherwise be unable to obtain. Communities may aim for partnerships when they can make more money from tree growing, harvesting or processing than alternatives would provide, but lack the means to exploit these advantages without the services that the company can provide.

Three country examples of company–community partnerships are described here, illustrating some of the different approaches that have been taken.

South Africa: outgrower schemes with livelihood benefits

In South Africa, outgrower schemes involve some 12,000 smallholder tree growers on about 27,000 hectares of land. The two schemes with the largest membership are operated by the country's biggest forestry companies, Sappi and Mondi. Smallholders grow trees with seedlings, credit, fertilizer and extension advice from companies who later buy the product for pulp. While outgrower timber only provides about 10 per cent of the two companies' mill throughput, and is more expensive per tonne than wood from other sources, it provides the fibre that would otherwise be unavailable because of land tenure constraints. This allows a volume of production to be reached that achieves economies of scale. Crucially, the schemes also provide companies with a progressive image at a time when the distribution of land rights in South Africa is being called into question. Two other outgrower schemes provide alternatives to the company schemes, one operated by a growers' union and the other by a cooperative. Apart from better representation, their respective advantages for members are shares in the downstream tannin factory and seeking the best prices for fibre.

For communities, outgrower schemes have contributed substantially to household income, providing participating households with about 20 per cent of the income needed to be just over the national 'abject poverty line', but they are yet to take households out of poverty. Small growers also face problems with opaque government policy and uncoordinated service provision from agencies of national and local government. Their associations lack the power to engage with the policies and institutions that affect their livelihoods. Nonetheless, outgrower schemes have had several positive impacts on communities' asset bases. Land rights have been secured and infrastructure has been developed in some areas. The schemes have been available to even the very poorest and most labour deficient of smallholders, because of the credit extended by companies, while non-landowners have benefited in some areas through employment as weeding, tending, harvesting or transport contractors to the landed smallholders.

Indonesia: third party roles and venture partnerships

In the Indonesian islands of Sumatra and Java, approximately 75 per cent of the total land area is classified as state forest land, falling under the jurisdiction of the Department of Forestry, which allocates logging and/or plantation rights to private companies. The government also has a central policy of promoting partnerships between companies and local smallholders or communities, with support from a Reforestation Fund that accrues from levies on logging.

One company that has greatly benefited from the Reforestation Fund has been PT Xylo Indah Pratama, a Sumatra-based company allied to Faber Castell. Unable to obtain sufficient raw materials for its pencil factory from its forest concession, the company used research and development to identify a local weedy species as a viable alternative. An outgrowing scheme based on 50–50 profit-sharing was established with smallholders who had unused land (mainly due to labour constraints). The scheme has not yet reached its first harvest, but both smallholders and company staff have already discovered just how much investment of time and effort is needed to maintain a workable relationship. Meanwhile, all of the financial risk is borne by the government through US$1 million in credit from the Reforestation Fund. The company will not be asked to meet repayments if its profits are insufficient, thus rendering the scheme vulnerable to changes in government policy.

All production forest in Java is under the control of PT Perhutani, a state-owned company that is in the process of being privatized. Perhutani allocates small plots within its teak plantations to local farmers for agroforestry, perpetuating the tumpang sari (taungya) system that has been in place for nearly 150 years. Farmers' opportunities to negotiate and influence decision-making within this scheme are limited, but recently innovations have appeared, albeit only on a localized scale. Local cooperatives have formed 'venture partnerships' with Perhutani, with contracts to manage tourism operations and other services around logging areas. These groups are showing the first signs of negotiating better deals with the company.

Ghana: social responsibility agreements

One outcome of the overhaul of forest policy in Ghana in the 1990s was a new regulation stipulating that companies tendering for timber cutting permits would be assessed in terms of their respect for the social and environmental values of local residents. Under the new law, which came into operation in 1998, logging companies are required to secure a 'Social Responsibility Agreement' with the customary owners of the land. This agreement follows a standard pattern, to include a code of conduct for the company's operations – guiding environmental, employment and cultural practices – and a statement of social obligations, which is a pledge of specific contributions to local development.

Ghana's Social Responsibility Agreements differ from many systems of corporate responsibility internationally in that each agreement must be fully negotiated with the local community. There is a strict procedure for developing an agreement with local representatives and the district forest office before submission to a central evaluation committee. While these agreements are still in their infancy, the policy itself has already provided useful lessons for other countries, where high-value timber is logged in community areas, in how to implement a fairly simple, cost-effective, accountable system to support sustainable and socially responsible logging.

Principles and success factors for good company–community forestry parnerships

The International Institute for Environment and Development (IIED), together with a range of collaborative research partners, examined 57 examples of company–community forestry partnership in 23 countries.[5] The examples covered a wide range of arrangements, including:

- farmer outgrower schemes to supplement company-grown fibre;
- community inter-cropping between company trees;
- local agreements around local timber and tourism concessions;
- joint ventures where communities put in land and labour;
- plantation protection services;
- access and compensation agreements.

Some key principles were identified by the study, which tend to foster better company–community forestry partnerships:

- A *formal and realistic contract* – legally valid but not over-complicated.
- *Security of contributions*, be they land, finance or labour, from both sides.
- *Shared understanding* of prospects and opportunities, as well as costs and risks.
- *Mechanisms for sharing decision-making* and information.
- A *joint work plan* – clear demarcation of each side's rights, responsibilities and expected rewards within an overall management framework.
- *Flexibility and space for negotiation*, including specific terms for review and revision.
- *Sustainable forest management practices*, in economic, social and environmental terms.
- *Extension and technical support*, as a regular rather than one-off service.
- *Procedures for conflict resolution* – arbitration, defection, termination and recourse.
- *Systems of accountability*, to the community (especially regarding benefit-sharing) and local government, and more widely to civil society.
- *Clear roles for third parties*, such as government, community development organizations and financing agents – drawing on their services and comparative advantage.
- *Integration with broader development plans* for the company, community, district and country.

Company–community forestry partnerships represent a very promising route for stakeholders to work together for SFM. There is much still to be learned, but it is clear that prospective partners should enter into such arrangements with their eyes wide open. Box 19.7 identifies some important success factors, with examples for improving partnerships over time.

---

5 J Mayers and S Vermeulen, *Company–Community Forestry Partnerships: From Raw Deals to Mutual Gains?* IIED, London, 2002, http://www.iied.org/forestry/pubs/psf.html#9132IIED

Box 19.7 **Some success factors and examples in improving company–community forest partnerships**

Companies	
Success factor	Example
• Staying abreast of the market – business innovation, paying market prices	• Several companies have moved into paying market prices for fibre in countries as far afield as India, Vanuatu, Guatemala and Zimbabwe
• Keeping ahead of legislation	• Companies going beyond basic social responsibility agreements have a business head start, eg in Ghana and Honduras
• Allowing sufficient time and resources to develop good working relations	• Long-term investment of staff time has paid off for companies in South Africa
• Being alert to broader economic, political and environmental change	• Companies are setting up outgrowing schemes in Pacific nations in anticipation of plantations eclipsing natural forests

Communities	
Success factor	Example
• Proactive planning to pre-empt the company in design and organization of key aspects of partnerships	• A village-level cooperative in Indonesia has negotiated a tourism contract on its own terms
• Business know-how and legal advice	• South African outgrowers have benefited from legal advice to improve the terms of their contracts
• Formation of a registered company	• Legal incorporation has paid off for communities in Papua New Guinea
• Action in second-best environments, in spite of risks	• Sometimes partnerships serve to secure shaky land rights, eg in Nicaragua

20 Conditions of Employment

SFM depends on work being carefully carried out with consideration for impacts on the environment and people as well as productivity. Well-paid, adequately trained and supervised, motivated and healthy employees will be able and willing to work productively and accurately. Human resources must be used as sustainably as forest resources for sustainable forest management to be achieved.

The International Labour Organization has described this as the Circle of Prosperity:

> 'If workers are in good health and well-equipped, if they are adequately trained, if work is properly organised and supervised and if there are adequate working conditions, workers will reach a high level of production, good wages and be able to afford good nutrition and housing, and enjoy a decent life. They will be able to improve their standard of living and work on a sustained basis over the day, the year and the whole working life. Young people will be attracted by employment in forestry.'[1]

In addition, forestry work often demands that workers are scattered through a forest. They must be capable and motivated to do a good job with minimal supervision often under difficult physical and social conditions. Well-trained workers need less supervision, permit more flexibility in work organization and are more content with their jobs.

WHAT IS REQUIRED?

Conditions of employment include all aspects of the relationship between the employer and employee. These include:

- wages and benefits;
- rights of representation and negotiation;
- health and safety provisions ;
- facilities and services for staff;
- training and skills development;
- opportunities for equity and profit-sharing.

Box 20.1 has some short case studies about conditions of employment. Box 20.2 gives an example of a process to improve working conditions in forestry. A number of ILO conventions are relevant to conditions of employment. Their application to forestry operations is covered in Appendix 3.3. Some tips for forest managers aiming to be good employers:

- adhere to all employment legislation and regulations;
- provide written contracts linked to payment system;
- recognize employees' rights to freely organize and negotiate through their chosen representatives;
- maintain good communication links between management and staff and establish procedures for negotiating and resolving complaints and disciplinary matters, with appeals and grievance procedures acceptable to all parties;
- pay fair wages (which equal or exceed the legal minimum wage) and benefits that are clearly related to performance level, responsibility and time worked. Bonus or incentive schemes also help

---

1 *Fitting the job to the Forest Worker*; An Illustrated Training Manual on Ergonomics. International Labour Organization, Geneva, 1992

Box 20.1 ## Conditions of employment – some case studies[2]

Management failing to listen to the needs of forest workers (Tanzania)

In 1992, a questionnaire survey was carried out on 80 randomly selected workers employed in three different forests. The results were rather disconcerting: only 12 workers showed positive attitudes towards work, 30 had a negative opinion on their job and the rest were indifferent. Reasons given were low pay, accident risks and lack of job security. Motivation was low due to lack of adequate tools, protective gear and work clothing; lack of wage incentives; lack of training and career development; lack of grievance procedures; tough working environment and lack of medical facilities. However, 38 per cent of the workers were strongly motivated by being able to cultivate farm plots. While some of the complaints can be explained by the general socio-economic conditions of the country, others are the result of management's ignorance or indifference and could easily be overcome through interventions at the local level.

Managers and forest workers benefit from improved communication (Fiji)

A survey of 176 logging workers in Fiji was carried out, to try to find ways of stabilizing the workforce, reducing labour turnover, and ensuring that training inputs were not lost through workers leaving the industry.

In response to the survey, the workers asked for better communication with managers and better planning of roads and landings, machine repair and time management. While wages were comparatively good, many workers expressed a preference for a fixed rate of pay plus a production bonus, rather than the existing hourly wage. Accident statistics were not available, but the survey served to highlight where protective equipment and safe working practices could be improved.

The survey demonstrated quite clearly that the workers took an interest in their jobs. They made numerous suggestions on how to improve working conditions and operational efficiency for their own and their company's benefit. All these questions were discussed in detail with the company managers, who were quite willing to act in order to reduce logging cost, increase productivity and stabilize the workforce.

Making use of workforce potential (Zimbabwe)

A survey of chainsaw operators in Zimbabwe was carried out, to establish what training would be necessary to ensure all operators were fully equipped to work well. Among other issues, the survey considered working conditions and wage levels, finding that contract labourers were much less well provided for than the direct employees. Practically all workers were interested in further training, and courses were arranged accordingly.

On the whole, the inquiry showed that the chainsaw operators had substantial work experience, enjoyed stable employment and were well motivated for work. However, there was scope for fuller use of their potential and the creation of conditions under which higher levels of quantity and quality of work, safety and job satisfaction could be achieved. What shocked management most was the observation that, daily, a tree feller wastes wood that is worth a month's wage. This clearly reflects the need for both training and an adequate wage level.

---

maintain good standards of work and productivity. Pay special attention to finding out which benefits are highly motivating, eg plots to cultivate land, or free education;

- ensure that men and women are paid equally for work of equal value and that there is no discrimination between employees on the basis of sex, race, political opinion, national extraction or social origin;

2 B Strehlke 'Asking forest workers about their job', *Commonwealth Forestry Review* 75(3), 1996

Box 20.2 **Globalization, contract labour and decent work in forestry – the example of Uruguay[3]**

One of the prominent features of globalization in forestry industries is the soaring share of timber produced from plantations of fast-growing tree species. Uruguay is one country offering very favourable conditions for such plantations and government policy has attracted large amounts of investment in them over the last decade.

The Uruguay government became concerned about the safety hazards in forestry as thousands of new and inexperienced workers entered the sector. This found favour with both employers and workers and regulations were developed after extensive consultation. During the discussions, it quickly became clear that the goal of safe work could not be pursued in isolation from other components of decent work. Social protection, the nature of employment contracts, skills and working and living conditions were all addressed in the regulations.

A particular characteristic of forest work is that most of it is carried out by contractors. In many countries some of the worst conditions in forest work are associated with contract labour. Government, employers and workers developed regulations clearly spelling out responsibilities in contracting situations and establishing a register of contractors with the Inspectorate of Labour. In an important step towards decent work in contracting businesses, a national association of forestry contractors was founded.

Work is under way to promote skills development and vocational training based on the regulations, and to strengthen the organizations of contractors and workers so that they can engage in effective social dialogue and collective bargaining. This strategy in Uruguay could well become an example of how globalization and decent work can go hand in hand.

- ensure reasonable working hours, not exceeding eight hours plus two hours break time;
- do not employ children (as defined by national legislation on child labour and compulsory schooling); do not employ workers under the age of 18 in hazardous or heavy work;
- provide basic first aid training and a first aid kit for all workers;
- ensure, as far as is reasonably practicable, that workplaces and equipment are safe;
- provide adequate protective equipment where necessary and training in its use;
- provide adequate medical facilities and evacuation procedures for accidents and emergencies;
- clarify responsibilities for accidents and their prevention in line with legislation and ensure all accidents are recorded and investigated;
- ensure occupational risks of diseases·such as malaria, digestive and respiratory disorders are minimized, are understood by staff, and are tackled with both preventative and curative measures;
- provide adequate housing facilities for employees and their families;
- ensure temporary camps have basic facilities for a healthy environment, especially for waste disposal and latrines;
- provide facilities in large, permanent camps for employees and their families, such as shops, schools, transport, and so on;
- provide appropriate training and skills development for employees at all levels, and equipment adequate to the task;
- consider incentives for productivity such as profit-sharing or share-options for employees, to provide them with a stake in the organization;
- do not use forced labour, including debt bondage.

---

3 P Poschen, *Social and Labour Dimensions of the Forestry and Wood Industries on the Move*. TMFWI/2001. International Labour Organization, Geneva, 2001. Available at www.gtz.de/forest_certification

21 Monitoring Social Impacts

Monitoring is an integral part of forest management and is fundamental to the continual improvement aimed for by the EMS. Just as the production and environmental effects of management need to be monitored, the forest organization's social effects need to be followed and, where appropriate, improved. Forest managers can keep track of how well they are tackling relevant social issues, by using approaches similar to those used in environmental and operational monitoring (see Chapter 17). This is also necessary because social values change over time.

Other stakeholders will also be keen to track the progress of actions which affect their interests. They are also likely to be 'monitoring' – although they may not express it in this way. If standards, objectives and targets have been developed and agreements made on social responsibilities, stakeholders involved in their development will want to check whether such targets are being achieved and agreements complied with.

If the number of stakeholders is small and/or the issues are fairly clear, a basic set of questions asked during informal communication can provide the necessary information to satisfy forest managers and other stakeholders that all interests are being looked after.

Where there is a greater range of stakeholders, forest values and/or operations – and where there is more contention in social issues – much can be gained from a systematic approach to monitoring and recording social impacts. This may require the involvement of independent social analysts or those with the skills to steer a process of participatory monitoring by the stakeholders themselves.

Organizations and businesses seeking to use monitoring to make progress towards sustainability are being encouraged to join an international undertaking called the Global Reporting Initiative.[1] The mission of this initiative is to develop and disseminate globally applicable sustainability guidelines for organizations incorporating stakeholder perspectives in their reporting on the economic, environmental and social dimensions of their activities.

Basic monitoring of social impacts, by the forest organization itself, enables it to check whether laws and regulations are being adhered to, whether agreements made with stakeholders are being complied with (by all parties), and whether key stakeholders are satisfied or not by the implementation of forest management. In this way, problems, as well as opportunities for improvement, can be identified.

Participatory monitoring, by the stakeholders themselves, enables the numerous benefits of collaboration (see Section 15.2) to be brought to bear in tracking progress:

---

1 Global Reporting Initiative. Sustainability Reporting Guidelines. www.globalreporting.org

- For the *forestry organization*, a participatory approach to monitoring helps in obtaining social information, and information on local perceptions of its impacts. It can help large forestry operations, which may have several collaborative arrangements to review, such as outgrowers schemes, access agreements, and NTFP collection agreements. It is also a way for the forestry organization to learn if it is tracking progress on stakeholders' own terms, and is making improvements in its management.
- It helps *other stakeholders* (such as forest authorities, trading partners and communities) to review their own objectives for entering into relations with a forestry organization, and the results being achieved.
- For *both the organization and other stakeholders*, monitoring social issues involves regular face-to-face contact between stakeholders. This, itself, is likely to bring the broader benefits of good communication, including understanding, mutual respect, trust and collaboration. It can also identify problems and opportunities early enough to agree responses.

Monitoring of social issues may involve a range of approaches, from a few basic questions which are asked from time to time by a representative of the forestry organization, to a process of participatory monitoring in which all stakeholders regularly interact and report to a liaison group.

There are two aspects to consider:

- social issues to monitor;
- approaches to monitoring social issues.

21.1 Social issues to monitor

Basic monitoring of social issues which should be carried out by any forestry organization with little or no external help or advice, should focus on the issues which have been identified as critical in:

 the organization's SFM policy (see Chapter 10)

 the (environmental and) social impact assessment (see Section 15.1)

 the organization's objectives and targets (see Chapter 11).

Some useful questions to consider when monitoring social issues include:

- Are the rights and tenure of stakeholders being recognized and respected by all other relevant stakeholders?
- Do forestry operations have the consent of those with rights to land and resources?
- In cases where tenure and rights have been diminished, are appropriate and agreed compensation mechanisms in place?
- Has compensation/mitigation been agreed and made available where stakeholders incur other costs?
- Are belief systems, cultural practices, sites and customs linked to the forest being recognized and respected by the forest organization?
- Are employment laws and legal conditions being adhered to?
- Are there agreements and effective procedures between employers and employees on conditions of service, work procedures, complaints and disciplinary matters?

• Are employees' rights to freely organize and negotiate being recognized?

In the case of participatory monitoring, the issues to be monitored should be defined and agreed by the stakeholders themselves.

Scale considerations!

Medium and larger-sized forest organizations are likely to be associated with a greater range of stakeholders, forest values and operations. Such situations demand more detailed monitoring of issues, and call for participatory monitoring.

21.2 Approaches to monitoring social issues

The most simple form of monitoring of social issues is to utilize a checklist of questions, like those listed above in Section 21.1. This can be used by one or more members of staff of the forestry organization specifically assigned the responsibility for attending to social issues. It is important that there is continuity in staff for this process. Through communication within the organization and with its stakeholders, the questions can be addressed.

In its simplest form, monitoring results can be recorded as simple 'yes' or 'no' answers. As experience increases, the checklists can be further developed to include more specific questions, scored answers, and weighting systems.

Participatory monitoring is a more ambitious approach. It involves developing a process for gathering, analysing and using information which stakeholders agree is relevant. In this process, stakeholders themselves define the indicators of good forest management, and they conduct the monitoring of these indicators. The results are then fed into the planning and review process or environmental management system.

Indicators are more likely to be 'real' to stakeholders if they have been generated by the stakeholders themselves: for example, *a man is poor if when he wants to go somewhere he has to borrow a bicycle*, may be more useful than a standard UN definition like, *a man is poor if his income is less than 50 per cent of the national average wage*.

To be useful to the company, however, these indicators should be easily related to the organization's objectives and targets. A stakeholder liaison group should therefore be encouraged to develop its indicators in light of full knowledge of these objectives and targets (see Chapter 11).

However, participatory monitoring is a relatively new approach – not yet routine in commercial forestry (although it is used in NGO and aid-supported community projects). The suggested framework for establishing a participatory monitoring process, shown below, is based on experiences in Brazil.

The main requirements of a framework for participatory monitoring are:

1 *Agree what the aim of monitoring is.* Identify who has a special interest in SFM or significant knowledge. Monitoring may be especially important if an experimental, or short-term pilot, approach has been adopted, and/or if some more sceptical stakeholders demand proof of certain outcomes.
2 *Identify what should be monitored.* This will tend to draw attention to those standards, objectives and targets which have social impacts, and to the environmental and economic impacts which stakeholders feel most strongly about.
3 *Identify indicators of progress (and failure).* The choice will depend on the availability of data and the ease with which it can be recorded. A good indicator will be Specific, Measurable, Action-oriented, Realistic, and Time-framed ('SMART'). For example: type and number of work-related accidents in a period; yield of NTFPs taken by agreement by communities; hours spent (by women) in collecting firewood each week; numbers of recreational visitors.
4 *Select the monitoring methods.* Methods chosen depend on the available time, skills, and resources. It

may be possible to find one method that can be used to assess several indicators at once. Some of the social objectives will need to be monitored using social tools, some of which are described in Box 19.5 (Social tools for working with stakeholders).

5 *Formulate the monitoring programme.* To avoid confusion, it is essential to clarify:
   - who will collect, compile, analyse, and disseminate information;
   - where monitoring is going to be carried out (for example, in which community, which area of forest);
   - which methods will be used;
   - when monitoring will occur (how often and which month/week/ day);
   - how costs will be borne (usually they will be borne by the forest organization, as part of its agreement with the stakeholder liaison committee).

6 *Finalize methods and train stakeholders in their use.* Test the methods and any tools used for measuring the indicators to ensure that they are both relevant and practical. Train those who will do the monitoring. Focal-point trainers, from local communities themselves, can become good general liaison people.

7 *Implement the monitoring programme.* It is important to be systematic in the collection of information in order to understand what changes are occurring, where and when.

8 *Deal with the information.* As far as possible, those who participated in information collection should take part in the analysis, to avoid misinterpretation of the information and findings. Sometimes a method allows for all stages to happen at once, for example a participatory map to assess if there are more villagers gathering a particular product from the forest this year as compared to last year. The group of villagers construct the map and identify which households have started gathering the product over the year – this is information collection. They add up how many these are – this is information compilation. They discuss why there are so few (or so many) and what can be done to improve the situation – this is information analysis.

9 *Document the analysis.* The content and format depend on the target audience. Transparency is strongly desirable.

10 *Use the results.* The forest manager will feed it into improved forest planning and management, through the EMS. The findings of the monitoring may be used by villagers to alter their activities, too. Both will use it to review any agreements.

22 Dealing with Erroneous Perceptions and Unrealistic Expectations

Any forestry organization which involves 'outsiders' and brings large-scale investment into an area inevitably raises questions and expectations among other stakeholders, particularly local people. These may be addressed by a forest organization, or they may be ignored. For reasons expressed throughout Part Five, the latter position is not tenable.

Most 'raised expectations' should be welcomed – a raising of expectations and hope is generally a driving force for people's motivation and capabilities for organization and collaboration.

WHAT IS REQUIRED?

Although forest organizations can have a major impact on people's welfare, they cannot be expected to directly achieve fundamental societal goals on their own.

As with other rural development projects, there is often an expectation that a forest organization can deliver general 'development'. This impression is often created by proponents of forestry, or the forest organization itself, if they have not taken the time to assess what the local implications of forest development really will be. When the (tacitly) promised, or assumed, benefits are not forthcoming, there can be immense frustration.

One solution is to ensure good communication from the beginning, and willingness to develop collaboration between stakeholders. This helps get expectations and realities out in the open, and allows realistic goals, standards and targets to be hammered out with key stakeholders.

It should be noted, however, that involving people in setting targets, and agreeing social objectives, involves their time and energy, which can leave them with high expectations for the results of their labours. If these are not followed up, it can lead to disillusionment and anger.

In addition to requiring stakeholders' time and energy, working with stakeholders, incorporating their needs and expectations in forest management, and implementing the wider requirements of SFM, creates considerable extra work for the forest manager. It can be daunting for the forest manager to attempt to meet the varying and potentially conflicting demands of their own organization and its various stakeholders, even if in the long term their expertise and abilities will be enhanced by the process.

Dealing with disputes over forest management may be a major problem for some forest managers

The introduction of SFM requirements, including dealing with social issues, into forest management is likely to be a step-by-step process. Managers may need to discuss, implement and monitor a realistic programme of improvement with stakeholders, allowing sufficient time to develop their abilities and without becoming overwhelmed by new demands. A phased approach to implementing SFM requirements (see Chapter 28) can provide a structured framework for implementing such a step-by-step process: it does not necessarily need to lead to certification, although this is one possible outcome.

Ways to deal with unrealistic expectations include:

- Hire/work with people who understand local issues.
- Ensure that staff who are in contact with local groups and government officials have a realistic picture of the scope of forest operations and associated impacts. Widely differing views among staff will lead to confused perceptions among other stakeholders.
- Respect different positions and keep communication channels open.
- Maximize regular face-to-face contact with stakeholders.
- Ensure continuity of approaches and stability of staff in positions which involve collaboration with stakeholders.
- Try to solve problems while they are still small.
- Give consultation processes plenty of time, and get stakeholders to focus on forestry-related priorities. Ensure it is understood by stakeholders that forestry cannot be expected to be the main vehicle for local development (an observation shared by most of the well-respected analysts of forestry today).
- In developing initiatives aimed at improved stakeholder benefits, start with small experiments in one area first, ensuring that stakeholders are part of the plan.
- Allow adaptation and flexibility in operations involving stakeholders.
- Build alliances with those who understand the real capacities of the forest organization to deliver social benefits.

Dealing with disputes and conflicts over forest management may be a major problem for some forest managers. Few managers have received training or feel confident to work with stakeholders to resolve disputes. Consideration should be given to training in conflict management, such as the short courses offered by the Centre for Rural Development and Training and the Regional Community Forestry Training Center for Asia and the Pacific (RECOFTC) (see Appendix 4.1).

Part Six

Forest Management Certification

Introduction to Part Six

Certification is the process of independent verification that forest management meets the requirements of a defined standard. Certification assesses the compliance of a particular forest management unit with the standard. When it is combined with an assessment of the chain of custody from the forest to the final product, an eco-label can be used to identify products from well-managed forests.

However, certification has developed beyond a market tool for the promotion of products of good forest management. In some countries, certification has been used as a means to implement government policies on sustainable forest management. In Mexico, for instance, forest law obliges the government to promote certification. Some importing country governments (eg the UK, Denmark and Germany) are also developing procurement policies aimed at promoting the purchase of wood products for government projects from sustainably managed sources. FSC certification is often mentioned as the prime example of mechanisms for identifying and verifying such sources.

Forest and chain of custody certification schemes are described in detail in the companion volume, *The Forest Certification Handbook*.[1] An overview of the main aspects of certification is provided here.

Credible certification schemes, including those for forest management, are usually made up of three elements, shown in Box Part 6-1:

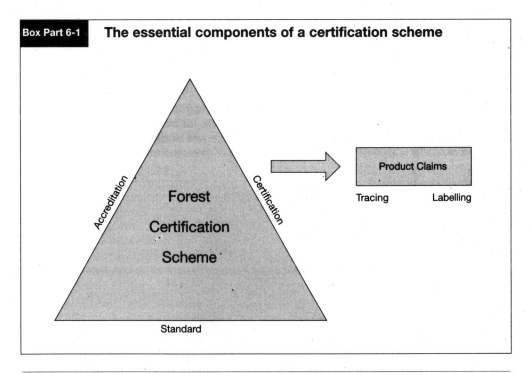

Box Part 6-1   The essential components of a certification scheme

Product Claims

Tracing   Labelling

Accreditation

Certification

Forest
Certification
Scheme

Standard

---

1 R Nussbaum and M Simula, *The Forest Certification Handbook*. 2nd edition. Earthscan, London, 2004

| Box Part 6-2 | **FSC-accredited certification bodies** |

Twelve certification bodies have been accredited, by the start of 2004, to carry out forest management certification and chain of custody assessments under the FSC scheme worldwide. In order to achieve accreditation under the FSC scheme, certification bodies must (among other requirements):

- Comply with the generic requirements of ISO/IEC Guide 65: 1996 (General requirements for bodies operating product certification systems).

- Develop a generic checklist based on the FSC P&C. Where there is an FSC-endorsed national or regional standard (see Section 3.5), certification bodies are required to use it for their assessments in that area.

- Document policies and procedures covering all steps of the certification process which meet the FSC's accreditation requirements.

- Post on their website a summary report of all the certification body's forest management or group certifications.

The FSC carries out monitoring visits to check both field and office procedures which are used by certification bodies.

- The *standard* sets out the requirements with which forest management must comply. The requirements of the FSC P&C and ITTO Guidelines have been discussed in Part Two.
- *Certification* is the process by which forest management is evaluated against the standard. Certification is carried out by an independent, third party called a certification body.
- *Accreditation* is the mechanism for ensuring that the certification body is independent, competent and produces reliable, replicable results. Accreditation sets out the rules for the certifiers and checks that they comply. Box Part 6-2 shows the accreditation requirements for FSC certification bodies.

In addition, if the certification is used for identifying products from certified forests and making claims about them, the scheme usually includes mechanisms for tracing and labelling products. *Chain of custody* certification is used for tracing forest products, and labelling rules define what claims a producer or retailer is allowed to make about the source of their products.

23 | Why Achieve Certification?

An important question all forest managers considering certification should ask themselves is whether or not to get their forest certified. This is not the same as asking whether or not to manage the forest sustainably, which should be the goal of every responsible manager. Certification is the process of verifying that forest management meets the requirements of a responsible forest management standard. It is only useful to confirm that you are implementing the requirements of a standard if there is a reason for doing so.

There are a number of reasons why a forest manager may decide to pursue certification. The most common are:

- customers are demanding certified products;
- there is the potential to use certification as a means of accessing new markets;
- an investor, donor or insurer is demanding certification as a conditionality on a loan, grant or insurance;
- the owners, shareholders or management see certification as a useful tool to achieve management goals or demonstrate best practice to stakeholders;
- pressure from governments as a condition for the allocation of forest management rights.

Each of these reasons for pursuing forest certification will have different implications for choosing which certification scheme to use. For example, there is little point in achieving forest certification under one scheme if customers are actually demanding a different type of certification. This is discussed further in Chapter 24 and Box 24.1.

24 Choosing a Certification Scheme

Over the past 10 years, a multitude of schemes for sustainable forest management certification have been developed, making it extremely difficult for the forest manager to know which SFM standards are most appropriate (see Box 24.1). There are subtle differences between the standards; meeting different standards may require different forest management practices.

The two major sets of international requirements relevant to tropical forests (FSC P&C and ITTO Guidelines) have been discussed in Part Two. The ITTO Guidelines are not directly used as a certification standard, but have been used to form the basis of several national and regional certification standards (see Section 3.5).

Given the plethora of different forest certification schemes that are available, it may be difficult for a forest organization to decide which scheme suits their needs. A great deal has been written about the merits of different forest certification schemes, not all of it objective in its analysis. Forest managers have often been left confused about which scheme they should opt for.

From the forest manager's point of view, there is no absolute answer to the question 'which scheme?' because it depends on the reason for achieving certification. The manager needs to consider why certification is needed and, based on that, decide which certification scheme is most suitable.

Sometimes this will be very clear because customers or investors will specify a particular scheme. At other times it will be less clear, and in this situation the best advice to managers is to ignore the politics and decide which scheme offers the greatest benefits. For example, if the reason for certification is to boost internal morale and improve relationships with local stakeholders, then it may be better to use a popular local scheme than an unpopular international one, even if the local scheme provides little market benefit.

In some cases, the best option may be to seek certification against two schemes simultaneously. This option is increasingly offered by certification bodies in countries where two or more schemes are operating. This may increase the total costs of certification, although certifying under two schemes simultaneously can reduce the costs of each one.

Any choice of a certification scheme must take into consideration the costs of certification under different schemes. In some countries, certification under a national scheme may entail a lower cost, because local accreditation and certification bodies operate more cheaply than international ones. However, a national certification scheme may not provide sufficient credibility for international markets.

The following sections offer some considerations relevant to deciding whether to pursue certification and choosing the most appropriate certification in a number of situations.

24.1 Customer Demand

One of the commonest reasons for seeking certification is that customers are demanding certified products. In this situation it is important to consider a number of questions, including:

- How many customers and what proportion of sales do they represent?
- Is the proportion likely to grow?
- Are they key customers, or likely to change supplier anyway?
- Which certification scheme(s) are they requesting and what does it cost?

Box 24.1 | **Forest certification schemes and their scopes (as of March 2004)**

FSC is a **global** forest certification scheme. The FSC P&C are interpreted at the national level into national/regional FSC Standards (see Box 3.4, Section 3.5). Where there is no national or regional standard, accredited certification bodies can carry out assessments using a generic standard based on the FSC P&C and including local indicators.

PEFC Council has endorsed 13 national schemes in **Europe** (Austria, Belgium, Czech Republic, Denmark, Finland, France, Germany, Latvia, Norway, Spain, Sweden, Switzerland, and the UK). Certification bodies are accredited through national accreditation bodies. Certfor in Chile and Cerflor in Brazil are the only two tropical members of PEFC; Certfor has submitted its scheme for PEFC endorsement.

MTCC (Malaysia) approves assessors directly to carry out assessments against the Malaysian Criteria, Indicators, Activities and Standards of Performance for Forest Management Certification (MC&I 2001). The MC&I have only been applied to the Permanent Forest Estate in **Malaysia**.

LEI (Indonesia) operates only in natural forest in **Indonesia**. Since 2000, LEI and FSC have operated under a Joint Certification Protocol which requires that joint assessments are undertaken and certificates can only be issued when forest management meets the requirements of both schemes. LEI approves assessors who carry out assessments against the LEI Standards; certification decisions are made by an expert panel appointed by LEI on the basis of the assessor's report.

SFI (Sustainable Forestry Initiative) was developed in the USA and has been applied to plantations and natural forests in the **USA and Canada**. The SFI Standard was revised in 2001 and the 2002–2004 SFI Standard and Verification Procedures were approved in 2002. Assessors are approved directly by the Sustainable Forestry Board (SFB) who developed and manage the SFI.

CSA (Canadian Standards Association) operates in **Canada**. The second edition of the standard (CAN/CSA Z809-02, A Sustainable Forest Management System: Requirements and Guidance) was approved in May 2003. The standard includes some process standard requirements as well as a requirement for a consultation process for setting locally determined performance requirements. Certification bodies are accredited by the Standards Council of Canada.

Certfor is a national forest certification initiative operating in **Chile**. It is led by an NGO, Fundacíon Chile, working in collaboration with the government agency for forest research (INFOR). Standards have been developed for plantation forests; further standards are being developed for native lenga forests. Four plantation operations have been certified (to February 2004), totalling 900,000 ha. The scheme has been submitted for endorsement by the PEFC.

Cerflor is a Brazilian government-approved scheme covering plantations in **Brazil**. The Cerflor standard for plantation forests was approved in January 2002; the natural forest standard is under development in 2004. In February 2004, one 50,000 ha plantation owned by INPACEL Agrofloresta Ltd (belonging to the International Paper Group) had been certified, while a second, Aracruz Cellulose, had begun the process.[1]

PAFCS (Pan-African Forest Certification Scheme) is at a very early stage of development and no certifications have yet been carried out under this scheme. The scheme is intended to cover all member countries of the African Timber Organisation (ATO). The standard (ATO/ITTO Principles, Criteria and Indicators for the Sustainable Management of African Natural Tropical Forests) was published in 2003.

---

1 FERN, *Footprints in the Forest: Current Practice and Future Challenges in Forest Certification*, February 2004. Available at www.fern.org

- Will improved sales cover the costs of certification?
- Will the customers assist with the costs of certification and under what conditions?

It is important to be accurate and realistic in assessing the demand for certification. If there is some resistance to certification in an organization then it may be tempting to underestimate demand. On the other hand, if there is enthusiasm for certification then it is easy to overestimate how many customers are looking for certified products. One way of assessing demand is to send customers a short questionnaire based on a series of simple questions such as:

- Which certification scheme are they interested in?
- Do they currently purchase any certified material? If so, what?
- What quantities and types of certified material would they like to purchase (a) now, or (b) in the future?

In addition to current customer demands, it may be important to consider what competitors are doing about certification and whether particular certification schemes are being promoted by local trade organizations. Achieving certification may allow an organization to expand their customer base, or may simply be necessary to maintain the market share they already command.

Considering these various demands, the organization needs to decide whether:

- customers (current and potential) are consistently demanding certification under a particular scheme;
- there is sufficient current demand to justify seeking certification under a particular scheme immediately;
- there is sufficient likelihood of future demand to justify starting preparations for certification immediately;
- there is no indication that certification is needed now or in the future.

24.2 Accessing New Markets

Certification is sometimes seen as a way to access new markets. International, niche markets demanding certification may appear to offer an attractive new opportunity to develop new products. However, careful consideration is needed before seeking certification as a means to enter new markets.

Developing new products is a costly and time-consuming business. It is essential to ensure that there really is a demand for the certified products at a price that can be attained. In the past, some producers have mistakenly assumed that demand for certified products was sufficiently strong so that other aspects of sales became less important. This has proved not to be true. Markets demanding certified products also demand high quality products, particular species, reliability of supply and competitive prices. In many cases, certified products enter niche markets, where quality is particularly important. Consumers who are paying a premium for the green credentials of their purchases are unlikely to be satisfied with poor quality goods. Certification is not a substitute for quality products.

Second, it is important to consider which certification schemes are acceptable to the potential new market. The forest and chain of custody certification must be achievable within the timescale for accessing the new market and affordable relative to the expected market value.

24.3 Investor, Donor or Insurer Demands

Increasingly investors, donors and insurers are using certification as a means of reducing their own risks and improving the quality of their investments. For example, some international banks which invest in forestry and agriculture have made commitments to use certification as a criterion in their due diligence procedures, to ensure their investments do not have negative social and environmental impacts.

Insurers may also use certification as a means of managing their risks, on the assumption that a certified well-managed forest is less likely to be damaged by social unrest or unforeseen environmental impacts. Donor funders of forestry projects also sometimes use certification as a means of evaluating and monitoring the success of the project, or may promote certification as a means of accessing markets. The World Bank/WWF Alliance has developed a system (The Questionnaire for Assessing the Comprehensiveness of Certification Schemes/Systems) for evaluating which forest certification schemes are consistent with their objectives and will receive their support.

If investors, donors or insurers demand that the forest should be certified, they usually will have decided which certification scheme they require. The forest manager needs to consider how best to achieve certification under that scheme. It may be possible to use a phased approach to certification (see Chapter 28), which allows the forest to progress through a number of verified steps to full certification.

24.4 Meeting Management Goals

Certification can provide a helpful framework for meeting internal management goals within an organization. If this is the main reason for seeking certification, the appropriate scheme will be the one whose standard best reflects the aims and values of the organization. However, there may be other considerations in choosing a certification scheme:

- If any of the other drivers for certification (described above) may become more important in the future, it might be worth choosing a certification scheme that will satisfy those needs at the same time. It may be possible to choose certification against one scheme, while still ensuring compatibility with another scheme, so that a second certification can be added at a later stage if it becomes desirable.
- Availability of tools, experience and expertise in applying particular schemes locally. It is very daunting to be the first operation in a country or region to apply a forest certification standard. The forest manager has to interpret the requirements and put them into practice in their forest without drawing on the help of local examples and expertise. Where there is local experience of certification under a particular scheme, there may be information and tools available from universities, NGOs, consultancies, trade organizations and certification bodies to help the forest manager implement the requirements.
- Accredited certification bodies are not available in all countries for all certification schemes. Local certification bodies are likely to be cheaper, more readily available to discuss the process, speak the local language and may be more culturally aware than international certifiers. This may be important in choosing a certification scheme.

24.5 Pressure from Governments

Certification is being used by a number of governments in both timber producing and timber importing countries as a means of identifying and encouraging good forestry practices. Governments have promoted or required certification by:

- making it a condition of awarding concessions or resource utilization rights. For example, the community concessions awarded by the Government of Guatemala in the Mayan Biosphere Reserve in the Peten Region, made forest certification a condition of the contract.[2]
- in some consuming countries, increasingly requiring their suppliers of wood products to ensure their sources are legal and sustainable. Certification is being used as a measure of sustainability.

Where governments are putting pressure on suppliers and forest managers to achieve certification, it is essential to know which certification schemes are acceptable to the government in question. Assistance may be available to help implement certification requirements: for example, USAID subsidized the initial costs of certification for community concessions in Guatemala.

Although there may be little choice about which certification scheme is acceptable to governments, there may be flexibility about the timescale for implementation. Some governments may support a phased approach to certification (see Chapter 28). Government procurement requirements being developed in the UK distinguish two categories of suppliers in their procurement. This starts with establishing the legality of the forest source, and promotes sourcing of forest products from 'legal and sustainable' sources, which may be demonstrated by certification.

---

2 C Soza, 'Forest certification in Guatemala' Annex 2 to the report 'Forest certification and communities: looking forward to the next decade', A Molnar, *Forest Trends*, January 2003

25 The Certification Process

Forest certification is applied at the operational forest management level. It is carried out for a particular forest management unit (FMU), which is defined as an area of forest under a single or common system of forest management. This might be a forest concession, a privately held area of forest land or a series of small areas of forest land owned by different people, but managed under a common system.

The certification process outlined here is based on the FSC certification scheme, which has been one of the most widely used to date, but certification against most other certification schemes will follow a similar pattern. However, differences do occur between schemes, particularly on the requirements for pre-assessment, stakeholder consultation, peer review and report writing.

Forest certification involves a number of stages, as outlined in Box 25.1. Each of the stages is discussed briefly below. A detailed account of the certification process, and the differences between schemes, is given in the companion volume, *The Forest Certification Handbook*.

25.1 Application and Proposal: Choosing your Certifier

Once a decision has been made to seek certification under a specific scheme, the first step is to choose a certification body (or certifier). Information on accredited certification bodies is usually available from the accreditation body or the certification scheme.

Box 25.1 **Outline of the certification process**

Application and Proposal

Pre-assessment

Close-out gaps

Stakeholder Consultation

Main assessment

Report and Peer Review

CERTIFICATION

Surveillance

On receipt of an application, certification bodies will send out a proposal, with an outline of their certification process and the costs. There is no obligation to proceed at this point. There may be several certification bodies operating in a region, in which case it is often useful to get a proposal from several certification bodies. There may be significant differences between certification bodies on costs, efficiency, local representation and reputation. If possible, it may be worth discussing with others who have been through the process their experiences of the certification process.

25.2 Pre-assessment: The Initial Visit

Once a certification body has been selected and a contract signed, there will normally be an initial pre-assessment (or scoping) visit to the forest. This short visit aims to allow:

- the forest manager and staff to meet the certifier and to ask any questions about the certification process.
- the certifier to familiarize themselves with the forest and to plan for the main assessment. The certifier may also use this opportunity to start identifying stakeholders who may be consulted prior to the main assessment.
- the certifier to discuss the requirements of the standard with the forest manager and see if there are any areas where forest management clearly does not comply. This is often referred to as 'identifying gaps'.

However, this is not an audit. The certifier does not check if the forest manager is accurately answering questions about compliance with the standard, so if problems are hidden, the gaps will not be identified at this stage.

Following the pre-assessment, the certification body sends the forest manager a confidential report, summarizing the findings and highlighting any gaps that need to be addressed before a main assessment.

25.3 Closing Out Gaps

If gaps are identified between current forest management practices and the requirements of the standards, these need to be addressed before moving on to a main assessment. This may take anywhere from a few days to many months, depending on what issues have to be addressed and what resources are available to address them.

Once the forest manager is satisfied that all outstanding issues have been addressed, the certifier can be contacted to arrange the main assessment.

25.4 Stakeholder Consultation

Stakeholder consultation aims to obtain people's positive and negative views and comments on the management of the FMU being assessed. Stakeholders may include neighbours and local communities, the forest department and local government, environmental and social NGOs, workers, employees, contractors and union representatives.

Some certification schemes require the certification body to undertake a consultation process with a range of stakeholders as part of the main assessment process. The FSC attaches much importance to stakeholder consultation. Certification bodies are required to contact key stakeholders at least a month before the main field assessment begins.

The certifier may consult by letter, on the telephone, through private or public meetings, by advertising in the local paper or any other way which seems appropriate. The number of stakeholders contacted and the method used for contacting them depend on the size and location of the forest and the type of stakeholders being contacted.

25.5 Main Assessment

The main assessment is usually carried out by an assessment team over a period of several days (depending on the size and complexity of the forest). The team is appointed by the certification body; some accreditation bodies specify the qualifications required for assessment team leaders and members. Usually the team comprises a team leader, who is familiar with the certification requirements, and a number of local experts (for example, specialists with detailed knowledge of legislation, forest operations, social, environmental or ecological issues).

The assessment usually starts with an opening meeting. This allows the team to be introduced to management staff and is an opportunity to discuss the main assessment process and logistics. During the main assessment, the team collect objective evidence to evaluate whether that forest management meets the requirements of the standard and to identify any areas where it does not. Three main types of objective evidence are examined:

- *Documentation*: a range of plans, documents and records are checked to ensure they meet the requirements of the standard and that adequate records are being kept.
- *Field observations*: a sample of sites in the forest is visited to check whether plans and procedures are being implemented. Compliance with legislation, regulations and local codes of practice is also checked.
- *Discussions and interviews*: the assessment team discusses with managers, staff, field workers, contractors and unions what they do and how they do it. Meetings with stakeholders are also used to gather information about forest management practices.

If areas are found where forest management does not comply with the standard, the assessment team records it as a corrective action request (CAR), also known as a condition. This sets out the details of the non-compliance and formally requires the forest organization to take action to address the problem. There are two types of CAR:

- Major CARs (*preconditions*) occur when there is complete failure to comply with a requirement of the standard or a systematic failure to implement plans and procedures. Major CARs must be addressed before certification can proceed.
- Minor CARs (*conditions*) occur when there is partial compliance with a requirement or a non-systematic failure to implement plans or procedures. Certification can still proceed but only on the condition that the issue is addressed within an agreed time.

The main assessment usually concludes with a closing meeting where the team report back on their findings. Each CAR is discussed and must be signed off by the forest management. The closing meeting provides an opportunity to point out if the team may have misunderstood an issue, or raised a CAR wrongly because they did not see some important evidence.

Any major CARs identified during the assessment are discussed at the closing meeting. These need to be addressed by the forest management before certification can proceed and the certification body needs to check that the action taken is adequate. If no major CARs are identified, the assessment team will usually recommend certification.

25.6 Report and Peer Review

Following the completion of the main assessment, the team leader, assisted by the team members, will write a report. The FSC requires that a summary of the report, including background information and the results of the assessment, is made public, so the report will include both the public summary and the more detailed assessment report. Other certification schemes do not require this type of public summary – only information on the forest or group (size, location, type etc) have to be made public.

Some schemes, including the FSC, require the report to be peer reviewed before any final certification decision is made. When the report is complete, it is sent to two or three independent specialists who have been selected as peer reviewers. As with the team members, the peer reviewers will be selected by the certification body, but you should be informed who they are and can tell your certifier if there is any problem.

The peer reviewers are asked to comment on whether, based on the information provided in the report, and their expert knowledge of the issues in the region and type of forest being assessed, the assessment seems to be adequate and the findings reasonable.

Any issues raised by peer reviewers must be responded to by the certifier and, if appropriate can lead to new or revised CARs.

25.7 Certification and Surveillance

Once all major CARs or preconditions have been adequately addressed, and any peer review comments have been responded to, the certification body will finally make a certification decision. This is normally done by a certification panel, which is independent of the assessment team (although the team leader may join the panel to answer questions and provide information).

Certification is also referred to as 'registration' by some schemes, particularly in north America. Usually this means exactly the same thing, so that reference to a 'registered forest organization' means the same as a 'certified forest organization'.

Certificates are usually valid for five years, but are conditional on the findings from annual surveillance visits, which the certification body will carry out to verify that the forest management in the forest organization or group scheme is continuing to comply with all the requirements of the standard.

Surveillance visits are usually like a shorter version of the main assessment, undertaken by a team with the same process being followed. The main things the surveillance team will focus on are:

- checking that any current minor CARs have been adequately addressed and can be 'closed out';
- checking that the actions taken to address previous major and minor CARs are continuing to be implemented and are successful in addressing the problem identified;
- visiting new sites or members – if new forest areas have been acquired, work has started in a new part of a concession or new members have joined a group scheme, the surveillance team are likely to want to visit,
- following up on any complaints or stakeholder comments received since the last visit;
- checking that any changes to the standard have been implemented – standards are usually revised at least once every five years and certificate holders usually then have 12 months to comply with any changes; the surveillance team will want to check this is being done (this is also the case if a forest was certified against a draft standard and the final standard has since been endorsed);
- assessing continued compliance against the standard.

As with the main assessment, any evidence that there is a non-compliance with the standard will result in corrective action requests.

26 Small and Low Intensity Managed Forests

It has become increasingly recognized that small forest enterprises and forests that are managed at a very low intensity (for example, for harvesting some non-timber forest products, or watershed protection) encounter particular problems in accessing and maintaining forest certification. These problems stem from a number of causes, in particular:[1]

- problems in understanding and meeting the requirements of the standard;
- the disproportionately high costs of certification in comparison to income derived from the forest.

In order to reduce these barriers to certification, some certification schemes have attempted to aim their standards and procedures at a level which is accessible to this target group. In particular, PEFC schemes in Europe which have largely been driven by Small Forest Owners' Associations have attempted to ensure access for small forest enterprises. This has been facilitated by the development of regional certification systems (see Box 26.1) and lower sampling requirements for certification. However, many environmental NGOs consider that this has led to PEFC certification being less rigorous than other certification schemes.[2]

Recognizing the difficulties faced by small forests and those managed at lower intensity, the FSC established its Small and Low Intensity Managed Forests (SLIMFs) Initiative in 2002. Through this initiative, the FSC has developed an adapted certification procedure for forests below a certain threshold size and those where timber production is at a low intensity (see Box 26.2).

Box 26.1 PEFC regional certification

Regional certification is the certification of forests within delimited geographical boundaries. The applicant must be an authorized organization for that region and must represent the forest owners or managers of more than 50 per cent of the forest area in the region. Commitment to forest certification may be made by a majority decision of a forest owners' organization on behalf of forest owners and managers in the region. Individual forest owners may withdraw from certification at any stage.

The national certification standard must contain criteria defined for the regional level as well as the forest management unit level. The applicant forest owners' organization is responsible for ensuring that forest owners/managers meet the certification requirements and for implementing the rules for regional certification. However, forest owners who are not members of the forest owners' organization and over whom it has no control, are apparently still covered by the regional certificate, leading to questions about how the forest owners' organization can ensure that they implement the certification requirements.

National schemes within the PEFC umbrella may determine the level of field sampling required to check that certification requirements are being implemented in the forest. Some schemes (such as PEFC Germany) have issued regional certificates without any field visits.

---

1 R Nussbaum, M Garforth, H Scrase, and M Wenban-Smith, 'An analysis of current FSC accreditation, certification and standard-setting procedures identifying elements which create constraints for small forest owners'. ProForest, 2001. Available at www.proforest.net

2 S Ozinga, *Footprints in the Forest: Current Practice and Future Challenges in Forest Certification*. FERN, February 2004. Available at www.fern.org

---

Box 26.2 | **Small and Low Intensity Managed Forests (SLIMFs): the FSC definition**

A forest can quality as a SLIMF if it is either 'small' or managed at a 'low intensity' (or both).

Small

- Forest management units of less than 100 hectares. FSC-accredited national initiatives may increase this limit to 1000 hectares.

Low intensity forest management units

- The rate of harvesting is less than 20 per cent of the mean annual increment (MAI) over the total production forest area of the FMU; *and*

  - EITHER the annual harvest from the total production area is less than 5000 cubic metres;

  - OR, the *average* annual harvest from the total production area is less than 5000 cubic metres during the validity of the certificate.

- The FMU is natural forest being managed for non-timber forest products; intensively managed non-timber forest product plantations are not considered as low intensity.

The streamlined procedures for smaller forests and those managed at a lower intensity are aimed at reducing the costs of certification. In particular, costs should be reduced for pre-assessments, reporting and peer reviews and the recurrent costs of monitoring visits.

In addition to the streamlined procedures, however, it will be essential to adapt forest certification standards to the reality of small forests and those managed at a lower intensity. In some countries, attempts have been made to incorporate these adaptations (see Box 26.3). In other countries, national or regional standards give little consideration to the needs of small and/or low intensity managed forests.

Box 26.3 | **Adapting standards for small forests: the example of the Regional FSC Forest Stewardship Standard for the Lake States – Central Hardwoods Region (USA), 2002**

Small forests in the region are defined in the Glossary as forests up to 5000 acres in size. The standard makes explicit reference to the criteria or indicators that small forest owners may be treated differently. For example:

Indicator 6.3.a.1. Forest owners or managers make management decisions using credible scientific information (eg site classification) and information on landscape patterns (eg land use/land cover, non-forest uses, habitat types); ecological characteristics of adjacent forested stands (eg age, productivity, health); species' requirements; and frequency, distribution, and intensity of natural disturbances.

Applicability Note: This indicator may apply only marginally to managers of small and mid-sized forest properties because of their limited ability to coordinate their activities with other owners within the landscape or to significantly maintain and/or improve landscape-scale vegetative patterns.

From 2005, FSC National Initiatives developing national or regional interpretations of the FSC P&C will be required to include adaptations in their standards, to take account of small and low intensity managed forests in their area. Standards will be required to be practical and cost-effective for use in small forests and those managed at low intensity. In particular, national initiatives will be required to provide alternative, appropriate indicators for:

- environmental impact assessment, protection of rare, threatened and endangered species, and protection of representative samples of existing ecosystems;
- management planning;
- monitoring and assessment;
- assessment of High Conservation Value Forests;
- restoration of natural forest cover and monitoring of on-site and off-site ecological and social impacts in plantations.

However, even with the development of streamlined certification procedures and adapted standards, many small forest enterprises find it cheaper and more convenient to approach certification as a group. In order to facilitate this process, group certification was developed and has become the most common mechanism for the certification of small forest enterprises.

27 Group Certification[1]

Managers of small forest enterprises often find it difficult to access forest certification. They may be in remote locations, with little access to information about certification requirements and procedures. In addition, costs of certification are disproportionately high for small forests because they do not benefit from economies of scale. Group certification can help to overcome some of these barriers.

27.1 Group and Resource Managers

Group certification allows a number of small forests to work together as a single entity, run by a group manager who can both provide information to the forest managers and organize a single certification assessment for the group, allowing them to benefit from the economies of scale of belonging to a larger unit. The 'group entity' might be an organization or association, a company or other legal entity. The group manager is responsible for ensuring that group members meet the requirements of the standard through their support to and monitoring of members' activities.

The group manager may be directly involved to a greater or lesser extent in members' forest management activities. A group manager may be responsible only for setting out the requirements of the group (how they should meet the standard), supporting and controlling the membership, and monitoring members' compliance with the group requirements. A group manager who is actually responsible for managing the forests on behalf of the owners is usually called a 'resource manager'. In practice, the group manager is often somewhere between the two extremes of 'group' or 'resource' manager, and may be directly involved in some aspects of forest management but not in others. Group certification can accommodate all these variations.

27.2 The Group Manager's Role

The group (or resource) manager is responsible for ensuring that the certification standard is met by all members of the group. In order to do so, the group manager needs to establish certain systems to manage and monitor the members: these systems are checked during the certification assessment. The following issues are normally considered when setting up a group management system:

- *Legal status* The group makes a contract for certification with the certification body and holds the certificate on the group's behalf. The group therefore needs to exist as a legal entity, such as a company, cooperative, association, NGO or community association.
- *Group management structure* The group is managed by a group manager, who is formally responsible for ensuring that the certification requirements are implemented. In a small group, group management is likely to be carried out by a single person. For larger groups, however, several people may be involved, sometimes working from different offices, or responsible for different management issues. It is essential that a single person is designated as the 'group manager' with overall control and that the management structure and responsibilities are clear.

---

1 For a full description of group certification see: R Nussbaum, 'Group certification for forests: a practical guide'. ProForest, 2002. Available at www.proforest.net

- *Who can join?* The type of members eligible to join the group needs to be defined at the outset. In practice, this is usually done by the group manager, according to their competencies and demand for membership. This may involve setting a minimum or maximum size forest, forest types which the manager is competent to deal with, the geographic area and locations that can be managed in practice, the number of members that can be managed, costs of joining and on-going fees.
- *Interpreting the standard* Forest certification standards often need interpretation before it is clear to a forest manager what they need to do in their specific situation. The group manager's role is to provide that interpretation for members, helping them decide what they need to do in practice in order to comply with the standard. This will generally involve developing a simplified version of the standard, possibly in a more accessible format, with specific prescriptions that apply to members' forest types and scale of operations.
- *Defining membership requirements* The group manager is responsible for ensuring that members meet the certification requirements. In order to do so, it is important to define clearly the requirements and procedures for joining the group, conditions for leaving the group and the procedures for expulsion from the group if requirements are not met. A formal membership agreement, signed by the group manager and the member, is necessary.
- *Consultation* FSC certification requires forest managers to maintain consultation with neighbours and affected parties. In group schemes, the group manager may take on some of the consultation role, for example, coordinating with local, regional or national stakeholders or government agencies on behalf of members.
- *Monitoring group members* One of the group manager's main tasks is to monitor members' activities and ensure that they comply with membership requirements. The group manager needs to develop a schedule for visiting members, a checklist for use during field monitoring, a reporting and feedback mechanism to ensure any problems are addressed and followed up, and systems for record-keeping.
- *Complaints* FSC certification requires that complaints are dealt with through an 'appropriate mechanism'. The group manager may need to define a formal mechanism for dealing with any complaints from stakeholders about members' activities, or with complaints from members about the group manager's activities.
- *Documents and records* The group manager needs to keep good records of members activities and may need to establish formal written procedures for managing the group.
- *Training and information* Depending on the information already available to group members, provision of training and information may be a major part of the group manager's role. The group manager is also responsible for providing public information about the group and its membership to stakeholders.
- *Controlling claims* The group manager is responsible for ensuring that group members and management only make claims in accordance with certification scheme rules (for example in publicity materials and product labels).

27.3 Group Certification in Practice

Group certification schemes aim to offer groups of small forests economies of scale by allowing them to undergo certification assessment as a single entity. The certification body carries out checks at two levels during the assessment (see Box 27.1):

- *an audit of the group management system*, to ensure that it is adequate to manage and monitor the group members' activities and to ensure that the standard is being met in members' forests;
- *an audit of a sample of group members* to check that the forests are being managed in accordance with the standard, and that the group manager's systems work in practice.

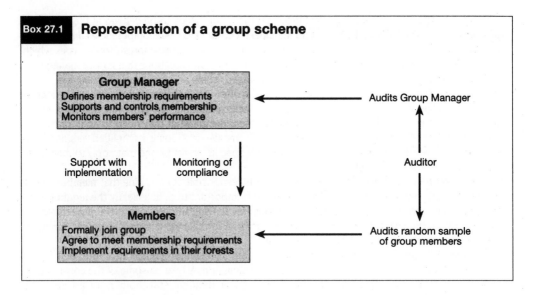

Box 27.1 Representation of a group scheme

Different certification schemes have slightly different requirements about the frequency with which both the group manager and the certification body need to monitor group members' activities. They also differ in the way in which weaknesses identified during an assessment in individual forests should be followed up at subsequent visits. However, all certification schemes will expect the group manager to ensure that there is a system in place to identify and correct any forestry activities which do not conform with the standard.

As a result of its SLIMFs Initiative, the FSC is developing streamlined procedures which reduce the burden of monitoring and reassessment on groups of small forests and those managed at a low intensity. In many small forests, activities such as harvesting, replanting or silviculture are carried out very infrequently and several years may pass during which very little management is actually carried out in the forest.

Both the group manager's and the certification body's sampling schedules need to be designed to take this into account. New FSC accreditation rules introduced in 2004, require that the group manager may base the frequency of their field monitoring of group members on the intensity of forest management activities. The group manager is required to visit every member's forest at least once during the validity of the certificate. Groups which are composed entirely of Small and Low Intensity Managed Forests (see Box 26.2 above) should be subject to field surveillance visits (of a sample of members) by the certification body a minimum of twice during the validity of the certificate, supplemented by documentary checks in intervening years.

28 Phased Approaches to Certification

Despite considerable improvements in forest management in many regions, many forests worldwide still do not meet certification requirements because forest management practices are well below the quality required by international standards. As a result, considerable work is required to implement all the changes needed to meet the standard. This presents huge challenges for forest managers, particularly if they have very limited staff and other resources to undertake the work.

The process of improvement is usually a lengthy one, requiring several years of concerted effort. It is often difficult for forest managers or external parties to assess clearly the progress made during this time since so many different activities are being undertaken.

Currently, there is no mechanism for providing an incentive for forest managers during the 'improvement period' to reward the investment being made. For example, buyers are unable to differentiate between products from forests in the transition to responsible forest management and certification and those from forests being poorly managed or even illegally harvested.

These problems can be overcome using a phased or stepped approach to the implementation of responsible forestry standards and certification:

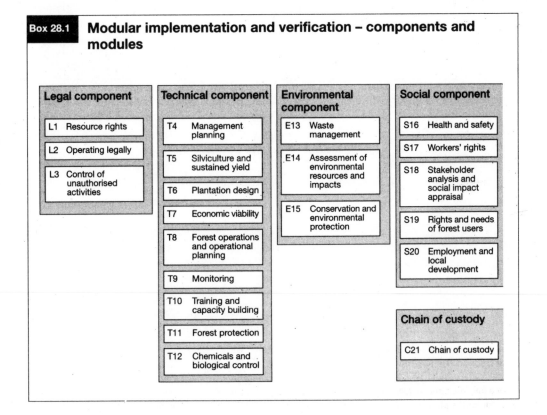

Box 28.1 **Modular implementation and verification – components and modules**

Legal component

- L1 Resource rights
- L2 Operating legally
- L3 Control of unauthorised activities

Technical component

- T4 Management planning
- T5 Silviculture and sustained yield
- T6 Plantation design
- T7 Economic viability
- T8 Forest operations and operational planning
- T9 Monitoring
- T10 Training and capacity building
- T11 Forest protection
- T12 Chemicals and biological control

Environmental component

- E13 Waste management
- E14 Assessment of environmental resources and impacts
- E15 Conservation and environmental protection

Social component

- S16 Health and safety
- S17 Workers' rights
- S18 Stakeholder analysis and social impact appraisal
- S19 Rights and needs of forest users
- S20 Employment and local development

Chain of custody

- C21 Chain of custody

- first, by undertaking one phase or step at a time, forest managers can focus limited resources on specific activities and make the process more manageable;
- second, it becomes much easier to assess the progress that has been made as one phase is completed and another begun;
- last, if the implementation of a phased approach is linked to a mechanism for the credible verification of progress, then a system for providing some type of incentive can be linked to the process.

Modular implementation and verification (MIV) provides a credible mechanism for delivering a phased approach which is practical, consistent and easy to communicate.[1] Modular implementation and verification works by dividing forestry standards into the parts that are commonly covered by all forestry standards: legal, technical, environmental and social components. Each of these components can be further subdivided into predetermined modules (see Box 28.1).

The division of forestry standards into practical modules allows each module to be addressed separately by the forest manager, implemented and verified on a step-by-step basis. As the requirements of the modules are predefined and clear, compliance with a particular module can be widely understood and communicated. However, at the same time it is possible to incorporate the specific requirements of any certification standard into the modular framework. A phased approach is compatible with, and can be used to work towards, certification under any appropriate certification scheme.

---

1 For a full description of MIV see: R Nussbaum, I Gray and S Higman, 'Modular Implementation and Verification (MIV): a toolkit for the phased application of forest management standards and certification'. ProForest, 2003. Available at www.proforest.net

Chain of Custody

A forest management certificate demonstrates that a forest is being managed in accordance with a defined level of practice. This may be useful for owners and managers of forests seeking investment, government or donor support or to provide stakeholders with a confirmation of good management practices. However, if the products of the certified forest are to be sold as products of a well-managed forest, it is essential that there is a mechanism for tracing the products to the forest of origin. This is the chain of custody. Inspection and verification of the chain-of-custody can result in chain of custody certification. Most certification schemes have developed rules for chain of custody certification that allow products from forests certified under their scheme to be traced to market and labelled as such.

Box 29.1 shows the rationale behind chain-of-custody inspection. Logs from a forest may pass through numerous manufacturing processes and different ownerships before reaching the final consumer. It may be sawn, peeled, chipped, broken down into fibre, divided into separate loads, combined with other types of wood or fibre, transported and stored. At any point in this chain, the products of certified forests could be mixed with those from uncertified forests. Chain-of-custody certification assesses the controls in place to ensure that mixing does not occur, or that it occurs in a controlled and permitted manner (see Section 29.1, Percentage-based Claims, below).

Ensuring a secure chain of custody usually relies on three different mechanisms:

- *segregation* keeps certified and uncertified materials physically separate, reducing chances of mixing;
- *identification* ensures that certified materials can be easily recognized at all processing stages, reducing the risk of accidental mixing;
- *documentation* of procedures, work instructions and records allows the origin of materials entering and leaving a process to be traced.

Taken together, these three mechanisms are usually adequate to trace the chain of custody of materials through a process. Adequate invoicing, records and procedures are also needed to trace certified products during transport, changes of ownership and storage.

29.1 Percentage-based Claims

For some products, it is possible to ensure that a final product is composed entirely of wood derived from certified forests. These are known as 100 per cent certified products.

However, for more complex chains it is extremely difficult to ensure that all the material input to the process is derived from certified forests. This may be the case, for example, where:

- the total area of certified forest in a region is insufficient to supply enough certified material;
- certified forests supplying a processor are dispersed at such a distance that requiring 100 per cent certified material would increase overall environmental impacts from transport over long distances;
- small and medium-sized forests, which have not achieved certification, supply a proportion of material to a processor and requiring 100 per cent certified material would exclude them from the market.

| Box 29.1 | Labelling – the rationale for inspection of the chain of custody |

Wood from a certified forest usually passes through several stages of manufacture before reaching the end user. A relatively simple example for wooden doors is shown below.

In the example above, the sawmill might well be handling both certified and uncertified logs. Therefore, the first stage in the chain-of-custody assessment is to ensure that these two sources are not mixed in the mill. To do this, the assessment team visits the mill and checks that certified and uncertified wood are always kept separate through a system of adequate segregation, identification and records. Some examples of the type of system required are shown below.

Activity	Segregation	Identification	Documentation
Log storage	Separate areas for certified and non-certified logs clearly signposted	Certified log ends painted red	Transport documents for all logs from certified forests kept on file with location in log yard added
Sawing	Batch process allows separation in time	Batches of certified logs tagged	Record book recording log numbers processed and batches of planks produced
Pre-shipment store	Separate areas for storage of certified and non-certified planks	Ends of certified planks painted red. Each batch labelled with a red numbered label	Store records and sales invoices record batch numbers

Following the assessment of the mill, a similar process would be required for the door factory to ensure that the chain is secure throughout.

Many certification schemes have therefore developed rules for making percentage-based claims about products. These essentially allow manufacturers or retailers to label products which contain less than 100 per cent certified material.

The rules for making percentage-based claims are complex and vary between certification schemes. Percentage-based claims are dealt with in more detail in the companion volume to this book *The Forest Certification Handbook*.

30 Other Certification-related Initiatives

There are two other major international initiatives related to certification, which may provide support, information and promotion of certification as a tool to improve forest management:

* World Bank WWF Alliance for Forest Conservation and Sustainable Use;
* WWF Global Forest and Trade Network (GFTN).

The World Bank WWF Alliance was formed with a specific target of achieving the independent certification of 200 million hectares of production forest by 2005, to be split between 100 million hectares in tropical countries and 100 million in temperate/boreal areas. In order to identify independent certification schemes which would contribute towards their target, the Alliance agreed nine principles and eleven criteria that it considers essential for certification. The FSC has been the only scheme which clearly meets the principles to date; an analysis of which other schemes might also meet the principles is under way in 2004.

The Alliance does not provide direct support for the certification of individual forest areas, but focuses on building the enabling conditions necessary for certification. It has therefore provided support for strengthening forest sector policy, developing national forest management standards, promoting Forest and Trade Networks (see below) and raising awareness about the problem of illegal logging.

The WWF *Global Forest and Trade Network* is WWF's initiative to eliminate illegal logging and improve the management of threatened and valuable forests. GFTN aims to facilitate trade links between companies committed to achieving and supporting responsible forestry practices. The GFTN is an affiliation of national and regional Forest and Trade Networks (FTNs), which include producer and consumer nations throughout Europe, the Americas, Russia, and Asia.[1]

Production-oriented FTNs, also called Producer Groups, are made up primarily of forest owners and managers who are working towards or have already achieved forest certification, as well as processors and manufacturers that are working towards the exclusion of illegal timber from their supply chain and greater trade in certified forest products.

Members of Producer Groups submit an action plan to their FTN showing how they will achieve certification or increase their trade in certified products. Members report to their FTN on progress against the action plan. This progress is also assessed periodically by an independent third party.

During the transition to certification, forest managers and owners can receive a variety of support services from the GFTN, including:

* information and technical assistance for improving forest management and achieving forest certification;
* market incentives for pursuing certification, though links with responsible purchasers in the network;
* support for advocacy efforts for changes to legislation and law enforcement.

---

1 More information about GFTN can be found at www.panda.org/forestandtrade/

Appendices

Appendix 1
The Standards

Appendix 1.1: The Forest Stewardship Council Principles and Criteria

FSC TRADEMARK. COPYRIGHT 1996 FOREST STEWARDSHIP COUNCIL.
A.C. (796/100)

Principles and Criteria For Forest Stewardship
Revised March 1996, edited October 1996
Revised January 1999

CONTENTS:

Introduction

It is widely accepted that forest resources and associated lands should be managed to meet the social, economic, ecological, cultural and spiritual needs of present and future generations. Furthermore, growing public awareness of forest destruction and degradation has led consumers to demand that their purchases of wood and other forest products will not contribute to this destruction but rather help to secure forest resources for the future. In response to these demands, certification and self-certification programs of wood products have proliferated in the marketplace.

The Forest Stewardship Council (FSC) is an international body which accredits certification organizations in order to guarantee the authenticity of their claims. In all cases the process of certification will be initiated voluntarily by forest owners and managers who request the services of a certification organization. The goal of FSC is to promote environmentally responsible, socially beneficial and economically viable management of the world's forests, by establishing a worldwide standard of recognized and respected Principles of Forest Stewardship.

The FSC's Principles and Criteria (P&C) apply to all tropical, temperate and boreal forests, as addressed in Principle 9 and the accompanying glossary. Many of these P&C apply also to plantations and partially replanted forests. More detailed standards for these and other vegetation types may be prepared at national and local levels. The P&C are to be incorporated into the evaluation systems and standards of all certification organizations seeking accreditation by FSC. While the P&C are mainly designed for forests managed for the production of wood products, they are also relevant, to varying degrees, to forests managed for non-timber products and other services. The P&C are a complete package to be considered as a whole, and their sequence does not represent an ordering of priority. This document shall be used in conjunction with the FSC's Statutes, Procedures for Accreditation and Guidelines for Certifiers.

FSC and FSC-accredited certification organizations will not insist on perfection in satisfying the P&C. However, major failures in any individual Principles will normally disqualify a candidate from certification, or will lead to decertification. These decisions will be taken by individual certifiers, and guided by the extent to which each Criterion is satisfied, and by the importance and consequences of failures. Some flexibility will be allowed to cope with local circumstances.

The scale and intensity of forest management operations, the uniqueness of the affected resources, and the relative ecological fragility of the forest will be considered in all certification assessments. Differences and difficulties of interpretation of the P&C will be addressed in national and local forest stewardship standards. These standards are to be developed in each country or region involved, and will be evaluated for purposes of certification, by certifiers and other involved and affected parties on a case by case basis. If necessary, FSC dispute resolution mechanisms may also be called upon during the course of assessment. More information and guidance about the certification and accreditation process is included in the FSC Statutes, Accreditation Procedures, and Guidelines for Certifiers.

The FSC P&C should be used in conjunction with national and international laws and regulations. FSC intends to complement, not supplant, other initiatives that support responsible forest management worldwide.

The FSC will conduct educational activities to increase public awareness of the importance of the following:

- improving forest management;
- incorporating the full costs of management and production into the price of forest products;
- promoting the highest and best use of forest resources;
- reducing damage and waste; and
- avoiding over-consumption and over-harvesting.

FSC will also provide guidance to policy makers on these issues, including improving forest management legislation and policies.

PRINCIPLE 1: COMPLIANCE WITH LAWS AND FSC PRINCIPLES

Forest management shall respect all applicable laws of the country in which they occur, and international treaties and agreements to which the country is a signatory, and comply with all FSC Principles and Criteria.

1.1 Forest management shall respect all national and local laws and administrative requirements.

1.2 All applicable and legally prescribed fees, royalties, taxes and other charges shall be paid.

1.3 In signatory countries, the provisions of all binding international agreements such as CITES, ILO Conventions, ITTA, and Convention on Biological Diversity, shall be respected.

1.4 Conflicts between laws, regulations and the FSC Principles and Criteria shall be evaluated for the purposes of certification, on a case by case basis, by the certifiers and the involved or affected parties.

1.5 Forest management areas should be protected from illegal harvesting, settlement and other unauthorized activities.

1.6 Forest managers shall demonstrate a long-term commitment to adhere to the FSC Principles and Criteria.

PRINCIPLE 2: TENURE AND USE RIGHTS AND RESPONSIBILITIES

Long-term tenure and use rights to the land and forest resources shall be clearly defined, documented and legally established.

2.1 Clear evidence of long-term forest use rights to the land (e.g. land title, customary rights, or lease agreements) shall be demonstrated.

2.2 Local communities with legal or customary tenure or use rights shall maintain control, to the extent necessary to protect their rights or resources, over forest operations unless they delegate control with free and informed consent to other agencies.

2.3 Appropriate mechanisms shall be employed to resolve disputes over tenure claims and use rights. The circumstances and status of any outstanding disputes will be explicitly considered in the certification evaluation. Disputes of substantial magnitude involving a significant number of interests will normally disqualify an operation from being certified.

PRINCIPLE 3: INDIGENOUS PEOPLES' RIGHTS

The legal and customary rights of indigenous peoples to own, use and manage their lands, territories, and resources shall be recognized and respected.

3.1 Indigenous peoples shall control forest management on their lands and territories unless they delegate control with free and informed consent to other agencies.

3.2 Forest management shall not threaten or diminish, either directly or indirectly, the resources or tenure rights of indigenous peoples.

3.3 Sites of special cultural, ecological, economic or religious significance to indigenous peoples shall be clearly identified in cooperation with such peoples, and recognized and protected by forest managers.

3.4 Indigenous peoples shall be compensated for the application of their traditional knowledge regarding the use of forest species or management systems in forest operations. This compensation shall be formally agreed upon with their free and informed consent before forest operations commence.

PRINCIPLE 4: COMMUNITY RELATIONS AND WORKERS' RIGHTS

Forest management operations shall maintain or enhance the long-term social and economic well-being of forest workers and local communities.

4.1 The communities within, or adjacent to, the forest management area should be given opportunities for employment, training, and other services.

4.2 Forest management should meet or exceed all applicable laws and/or regulations covering health and safety of employees and their families.

4.3 The rights of workers to organize and voluntarily negotiate with their employers shall be guaranteed as outlined in Conventions 87 and 98 of the International Labour Organization (ILO).

4.4 Management planning and operations shall incorporate the results of evaluations of social impact. Consultations shall be maintained with people and groups directly affected by management operations.

4.5 Appropriate mechanisms shall be employed for resolving grievances and for providing fair compensation in the case of loss or damage affecting the legal or customary rights, property, resources, or livelihoods of local peoples. Measures shall be taken to avoid such loss or damage.

PRINCIPLE 5: BENEFITS FROM THE FOREST

Forest management operations shall encourage the efficient use of the forest's multiple products and services to ensure economic viability and a wide range of environmental and social benefits.

5.1 Forest management should strive toward economic viability, while taking into account the full environmental, social, and operational costs of production, and ensuring the investments necessary to maintain the ecological productivity of the forest.

5.2 Forest management and marketing operations should encourage the optimal use and local processing of the forest's diversity of products.

5.3 Forest management should minimize waste associated with harvesting and on-site processing operations and avoid damage to other forest resources.

5.4 Forest management should strive to strengthen and diversify the local economy, avoiding dependence on a single forest product.

5.5 Forest management operations shall recognize, maintain, and, where appropriate, enhance the value of forest services and resources such as watersheds and fisheries. The rate of harvest of forest products shall not exceed levels which can be permanently sustained.

PRINCIPLE 6: ENVIRONMENTAL IMPACT

Forest management shall conserve biological diversity and its associated values, water resources, soils, and unique and fragile ecosystems and landscapes, and, by so doing, maintain the ecological functions and the integrity of the forest.

6.1 Assessment of environmental impacts shall be completed – appropriate to the scale, intensity of forest management and the uniqueness of the affected resources – and adequately integrated into management systems. Assessments shall include landscape level considerations as well as the impacts of on-site processing facilities. Environmental impacts shall be assessed prior to commencement of site-disturbing operations.

6.2 Safeguards shall exist which protect rare, threatened and endangered species and their habitats (e.g., nesting and feeding areas). Conservation zones and protection areas shall be established, appropriate to the scale and intensity of forest management and the uniqueness of the affected resources. Inappropriate hunting, fishing, trapping and collecting shall be controlled.

6.3 Ecological functions and values shall be maintained intact, enhanced, or restored, including:
   a) Forest regeneration and succession.
   b) Genetic, species, and ecosystem diversity.
   c) Natural cycles that affect the productivity of the forest ecosystem.

6.4 Representative samples of existing ecosystems within the landscape shall be protected in their natural state and recorded on maps, appropriate to the scale and intensity of operations and the uniqueness of the affected resources.

6.5 Written guidelines shall be prepared and implemented to: control erosion; minimize forest damage during harvesting, road construction, and all other mechanical disturbances; and protect water resources.

6.6 Management systems shall promote the development and adoption of environmentally friendly non-chemical methods of pest management and strive to avoid the use of chemical pesticides. World Health Organization Type 1A and 1B and chlorinated hydrocarbon pesticides; pesticides that are persistent, toxic or whose derivatives remain biologically active and accumulate in the food chain beyond their intended use; as well as any pesticides banned by international agreement, shall be prohibited. If chemicals are used, proper equipment and training shall be provided to minimize health and environmental risks.

6.7 Chemicals, containers, liquid and solid non-organic wastes including fuel and oil shall be disposed of in an environmentally appropriate manner at off-site locations.

6.8 Use of biological control agents shall be documented, minimized, monitored and strictly controlled in accordance with national laws and internationally accepted scientific protocols.

Use of genetically modified organisms shall be prohibited.

6.9 The use of exotic species shall be carefully controlled and actively monitored to avoid adverse ecological impacts.

6.10 Forest conversion to plantations or non-forest land shall not occur, except in circumstances where conversion:

a) entails a very limited portion of the Forest Management Unit; and

b) does not occur on High Conservation Value Forest areas; and

c) will enable clear, substantial, additional, secure, long-term conservation benefits across the Forest Management Unit

PRINCIPLE 7: MANAGEMENT PLAN

A management plan – appropriate to the scale and intensity of the operations – shall be written, implemented, and kept up to date. The long-term objectives of management, and the means of achieving them, shall be clearly stated.

7.1 The management plan and supporting documents shall provide:

a) Management objectives.

b) Description of the forest resources to be managed, environmental limitations, land use and ownership status, socio-economic conditions, and a profile of adjacent lands.

c) Description of silvicultural and/or other management system, based on the ecology of the forest in question and information gathered through resource inventories.

d) Rationale for rate of annual harvest and species selection.

e) Provisions for monitoring of forest growth and dynamics.

f) Environmental safeguards based on environmental assessments.

g) Plans for the identification and protection of rare, threatened and endangered species.

h) Maps describing the forest resource base including protected areas, planned management activities and land ownership.

i) Description and justification of harvesting techniques and equipment to be used.

7.2 The management plan shall be periodically revised to incorporate the results of monitoring or new scientific and technical information, as well as to respond to changing environmental, social and economic circumstances.

7.3 Forest workers shall receive adequate training and supervision to ensure proper implementation of the management plan.

7.4 While respecting the confidentiality of information, forest managers shall make publicly available a summary of the primary elements of the management plan, including those listed in Criterion 7.1.

PRINCIPLE 8: MONITORING AND ASSESSMENT

Monitoring shall be conducted – appropriate to the scale and intensity of forest management – to assess the condition of the forest, yields of forest products, chain of custody, management activities and their social and environmental impacts.

8.1 The frequency and intensity of monitoring should be determined by the scale and intensity of forest management operations as well as the relative complexity and fragility of the affected environment. Monitoring procedures should be consistent and replicable over time to allow comparison of results and assessment of change.

8.2 Forest management should include the research and data collection needed to monitor, at a minimum, the following indicators:

a) Yield of all forest products harvested.

b) Growth rates, regeneration and condition of the forest.

c) Composition and observed changes in the flora and fauna.

d) Environmental and social impacts of harvesting and other operations.

e)   Costs, productivity, and efficiency of forest management.

8.3   Documentation shall be provided by the forest manager to enable monitoring and certifying organizations to trace each forest product from its origin, a process known as the 'chain of custody.'

8.4   The results of monitoring shall be incorporated into the implementation and revision of the management plan.

8.5   While respecting the confidentiality of information, forest managers shall make publicly available a summary of the results of monitoring indicators, including those listed in Criterion 8.2.

PRINCIPLE 9: MAINTENANCE OF HIGH CONSERVATION VALUE FORESTS

Management activities in High Conservation Value Forests shall maintain or enhance the attributes which define such forests. Decisions regarding High Conservation Value Forests shall always be considered in the context of a precautionary approach.

Criteria

9.1   Assessment to determine the presence of the attributes consistent with High Conservation Value Forests will be completed, appropriate to scale and intensity of forest management.

9.2   The consultative portion of the certification process must place emphasis on the identified conservation attributes, and options for the maintenance thereof.

9.3   The management plan shall include and implement specific measures that ensure the maintenance and/or enhancement of the applicable conservation attributes consistent with the precautionary approach. These measures shall be specifically included in the publicly available management plan summary.

9.4   Annual monitoring shall be conducted to assess the effectiveness of the measures employed to maintain or enhance the applicable conservation attributes.

PRINCIPLE 10: PLANTATIONS

Plantations shall be planned and managed in accordance with Principles and Criteria 1–9, and Principle 10 and its Criteria. While plantations can provide an array of social and economic benefits, and can contribute to satisfying the world's needs for forest products, they should complement the management of, reduce pressures on, and promote the restoration and conservation of natural forests.

10.1 The management objectives of the plantation, including natural forest conservation and restoration objectives, shall be explicitly stated in the management plan, and clearly demonstrated in the implementation of the plan.

10.2 The design and layout of plantations should promote the protection, restoration and conservation of natural forests, and not increase pressures on natural forests. Wildlife corridors, streamside zones and a mosaic of stands of different ages and rotation periods, shall be used in the layout of the plantation, consistent with the scale of the operation. The scale and layout of plantation blocks shall be consistent with the patterns of forest stands found within the natural landscape.

10.3 Diversity in the composition of plantations is preferred, so as to enhance economic, ecological and social stability. Such diversity may include the size and spatial distribution of management units within the landscape, number and genetic composition of species, age classes and structures.

10.4 The selection of species for planting shall be based on their overall suitability for the site and their appropriateness to the management objectives. In order to enhance the conservation of biological diversity, native species are preferred over exotic species in the establishment of plantations and the restoration of degraded ecosystems. Exotic species, which shall be used only when their performance is greater than that of native species, shall be carefully monitored to detect unusual mortality, disease, or insect outbreaks and adverse ecological impacts.

10.5 A proportion of the overall forest management area, appropriate to the scale of the plantation and to be determined in regional standards, shall be managed so as to restore the site to a natural forest cover.

10.6 Measures shall be taken to maintain or improve soil structure, fertility, and biological activity. The techniques and rate of harvesting, road and trail construction and maintenance, and the choice of species shall not result in long-term soil degradation or adverse impacts on water quality, quantity or substantial deviation from stream course drainage patterns.

10.7 Measures shall be taken to prevent and minimize outbreaks of pests, diseases, fire and invasive plant introductions. Integrated pest management shall form an essential part of the management plan, with primary reliance on prevention and biological control methods rather than chemical pesticides and fertilizers. Plantation management should make every effort to move away from chemical pesticides and fertilizers, including their use in nurseries. The use of chemicals is also covered in Criteria 6.6 and 6.7.

10.8 Appropriate to the scale and diversity of the operation, monitoring of plantations shall include regular assessment of potential on-site and off-site ecological and social impacts, (e.g. natural regeneration, effects on water resources and soil fertility, and impacts on local welfare and social well-being), in addition to those elements addressed in Principles 8, 6 and 4. No species should be planted on a large scale until local trials and/or experience have shown that they are ecologically well-adapted to the site, are not invasive, and do not have significant negative ecological impacts on other ecosystems. Special attention will be paid to social issues of land acquisition for plantations, especially the protection of local rights of ownership, use or access.

10.9 Plantations established in areas converted from natural forests after November 1994 normally shall not qualify for certification. Certification may be allowed in circumstances where sufficient evidence is submitted to the certification body that the manager/owner is not responsible directly or indirectly for such conversion.

Principles 1–9 were ratified by the FSC Founding Members and Board of Directors in September 1994. Principle 10 was ratified by the FSC Members and Board of Directors in February 1996. The revision of Principle 9 and the addition of Criteria 6.10 and 10.9 were ratified by the FSC Members and Board of Directors in January 1999.

Appendix 1.2: ITTO Guidelines for the Sustainable Management of Natural Tropical Forests

ITTO GUIDELINES FOR THE SUSTAINABLE MANAGEMENT OF
NATURAL TROPICAL FORESTS

ITTO POLICY DEVELOPMENT SERIES 1

International Tropical Timber Organization (ITTO)
International Organizations Center, 5th Floor
Pacifico-Yokohama, 1–1–1, Minato-Mirai
Nishi-ku, Yokohama 220, Japan
July 1992

CONTENTS

4   Socio-Economic and Financial Aspects
    4.1   Relations with Local Populations
    4.2   Economics, Incentives, Taxation

1 Introduction

These guidelines contain a set of principles which constitutes the international reference standard established by ITTO for the development of more specific guidelines, at the national level, for sustainable management of natural tropical forests for timber production. The development, application and enforcement of national guidelines based on this standard are matters for national decision by individual timber producing countries.

The present reference standard is based on the report of a Working Group established in accordance with Council Decision 3 (VII). It has been elaborated on the basis of the Terms of Reference provided by the programme of work for ITTO in the field of reforestation and forest management for the year 1990, endorsed by the Council at its Seventh Session in November 1989. The Working Group Report was tabled at the Sixth Session of the Permanent Committee on Reforestation and Forest Management and adopted by the Eighth Session of the Council in May 1990. This initiative of ITTO refers to objective 1(h) of the International Tropical Timber Agreement, 1983: 'To encourage the development of national policies aimed at sustainable utilization and conservation of tropical forests and their genetic resources, and at maintaining the ecological balance in the regions concerned.'

The adoption by ITTO and its member countries of international guidelines that constitute a reference standard for sustainable management of natural forests is in the best interest of all producer and consumer countries which are concerned with the efficient and sustainable development of the tropical forest resources and forest-based industries.

ITTO attaches high priority to the definition of the essential principles and associated actions which should serve to guide the development of national guidelines in each country, in order that they may conform to the international reference standard agreed within the Organization. The Organization also gives high priority to assist member countries, which may need and request such assistance, to obtain such outside technical and financial help as they may require to develop their own national guidelines.

The ITTO guidelines are presented in the form of principles and possible actions covering considerations ranging from general policy to forestry operations aspects. Where available, examples of elements for possible inclusion in national and operational guidelines are given in appendices.

2 Policy and Legislation

2.1 FOREST POLICY

Principle 1. A strong and continued political commitment at the highest level is indispensable for sustainable forest management to succeed.

> *Possible action* 1. A national land use policy aiming at the sustainable use of all natural resources, including the establishment of a permanent forest base, should be developed and adopted.

> *Possible action* 2. A national forest policy forming an integral part of the national land use policy, assuring a balanced use of forests, should be formulated by means of a process seeking the consensus of all the actors involved: government, local population and the private sector.

Possible action 3. The organization of seminars for discussing policy, involving the above-mentioned interest groups.

Considerations in deciding a forest policy include the present proportion of land under forest; needs of protection and conservation of biological diversity (see Appendix 1); needs and aspirations of present and future generations of the population; the place of forestry in national economic planning; the various objectives of forest policy and relative importance of these; the amount of public and private forests.

Principle 2. An agreed forest policy should be supported by appropriate legislation which should, in turn, be in harmony with laws concerning related sectors.

Possible action 4. Laws and regulations at appropriate national and local government levels should be enacted, or revised as needed, to support the established forest policy, in harmony with policies, laws and regulations in related sectors.

Principle 3. There should be a mechanism for regular revision of policy in the light of new circumstances and/or availability of new information.

Possible action 5. Provision of adequate funds for research and monitoring to allow updating of policies.

Possible action 6. Research on evaluation of full economic benefits (total of marketed and non-marketed goods and services), provided by forests primarily managed for timber production, to enable foresters to better state the case for natural forest management for sustained timber production.

2.2 NATIONAL FOREST INVENTORY

Principle 4. A national forest inventory should establish the importance of all forests, independent of their ownership status, for the purposes identified in section 2.1 (see also Appendix 2).

Principle 5. There should be flexible provisions for such inventories to be broadened to include information not previously covered, if and when the need and opportunity for such additional information arises.

2.3 PERMANENT FOREST ESTATE

Principle 6. Certain categories of land, whether public or private, need to be kept under permanent forest cover to secure their optimal contribution to national development.

Principle 7. The different categories of land to be kept under permanent forest are (see also Appendix 1): land to be protected; land for nature conservation; land for production of timber and other forest products; land intended to fulfill combinations of these objectives.

Possible action 7. To identify, survey and boundary mark the various categories of the Permanent Forest Estate, in consultation with surrounding populations, taking into account their present and future needs for agricultural land and their customary use of the forest.

Principle 8. Land destined for conversion to other uses (agriculture, mines, etc.), and any land for which the final use is uncertain, should be kept under managed forest until the need for clearing arises.

2.4 FOREST OWNERSHIP

Principle 9. The principles and recommendations for implementation in these guidelines apply equally strictly to national forests and privately owned or customarily held forests.

2.5 NATIONAL FOREST SERVICE

Principle 10. There should be a national agency capable of managing the government forest estate, and assisting in the management of private and customarily held forests, according to the objectives laid in the national forest policy.

> *Possible action* 8. Provide for such a national agency.

3 Forest Management

Principle 11. Forests set aside for timber production are able to fulfill other important objectives, such as environmental protection and, to a varying extent, conservation of species and ecosystems. These multiple uses should be safeguarded by the application of the environmental standards, spelled out below, to all forest operations.

3.1 PLANNING

Principle 12. Proper planning, at national, forest management unit and operational levels, reduces economic and environmental costs and is therefore an essential component of long-term sustainable forest management.

> *Possible action* 9. To make adequate provision for forest management planning capacity at all administrative levels.

3.1.1 Static and dynamic inventory

Principle 13. The forests set aside for timber production should be the subject of a more detailed inventory to allow for planning of forest management and timber harvesting operations. The question of type and quantity of data to be gathered should be the subject of cost-benefit analysis.

> *Possible action* 10. Carry out inventories, concentrating on quantities of timber of currently and potentially commercial tree species of the forest for future timber production (see also Appendix 2).

> *Possible action* 11. To establish representative series of permanent sample plots.

3.1.2 Setting of management objectives

Principle 14. Management objectives should be set rationally for each forest management unit. Formulation of objectives should allow the forest manager to respond flexibly to present and future variations in physical, biological and socio-economic circumstances, keeping in mind the overall objectives of sustainability.

Principle 15. The size of each production forest management unit should preferably be a function of felling cycle, the average harvested volume per ha and annual timber outturn target of the operating agency (state forest enterprise, concessionaire, etc.)

3.1.3 Choice of silvicultural concept

Principle 16. The choice of silvicultural concept should be aimed at sustained yield at minimum cost, enabling harvesting now and in the future, while respecting recognized secondary objectives.

> *Possible action* 12. To gather information which provides the basis for rational choice of silvicultural practices, such as inventories and measurements from growth and yield plots, as well as data on market demand for various end uses of timber products. A true progressive silvicultural system should be developed by gradually improving on these practices as better information becomes available. The harvesting intensity and the design of harvesting pattern should be integral parts of the silvicultural concept.

3.1.4 Yield regulation, Annual Allowable Cut (AAC)

Principle 17. In order to ensure a sustained production of timber from each forest management unit, a reliable method for controlling timber yield should be adopted.

> *Possible action* 13. The Annual Allowable Cut (AAC) should be set conservatively in the case of absence of reliable data on the regeneration and growth dynamics of tree species, especially with regard to diameter increment and response to the effect of logging on trees and soil. This applies both to tree species which, under current market conditions, are desirable or which have the potential to become commercially attractive in the future, recognizing that domestic and world markets for forest produce are under very dynamic development. In practice, this will often mean conservative setting of rotation length, felling cycle and girth limits. As and when permanent sample lots begin to yield more reliable information about dynamics of desirable species, a reassessment of AAC should be considered.

> *Possible action* 14. To make provision for regular review of AAC (5-yearly) in order to take account of replacement of original forests by managed forests and the transfer of conversion forest to other uses. In the longer term, stand modelling should be introduced to assure efficient and responsible yield regulation.

3.1.5 Management inventory and mapping

Principle 18. A management inventory supported by a detailed map is indispensable to the preparation of working plans for each forest management unit.

> *Possible action* 15. Management inventory and mapping should be carried out.

3.1.6 Preparation of working plans

Principle 19. Working Plans should guarantee the respect of environmental standards in field operations.

> *Possible action* 16. Preparation of Working Plans including the following details (see also 3.2.3):
> * sequence of annual harvesting areas and allocation of all-weather and dry-weather areas;
> * areas to be excluded from harvesting;
> * road and extraction track layout;
> * details of marking, harvesting, post-harvesting inventory; silvicultural treatments;
> * fire management plan.

3.1.7 Enviromental impact assessment

Principle 20. Forest management operations can have important positive or negative environmental consequences, both in the forest itself and outside (transboundary effects). These consequences should be assessed in advance of operations to ensure overall sustainability.

> *Possible action* 17. Specify conditions under which an Environmental Impact Assessment (EIA) should be required.

Possible action 18. Design EIA procedure and provide for qualified staff to carry our EIAs.

3.2 HARVESTING

Principle 21. Harvesting operations should fit into the silvicultural concept, and may, if they are well planned and executed, help to provide conditions for increased increment and for successful regeneration. Efficiency and sustainability of forest management depend to a large extent on the quality of harvesting operations. Inadequately executed harvesting operations can have far-reaching negative impacts on the environment, such as erosion, pollution, habitat disruption and reduction of biological diversity, and may jeopardize the implementation of the silvicultural concept.

3.2.1 Pre-harvest prescriptions

Principle 22. Pre-harvest prescriptions are important to minimize logging damage to the residual stand, to reduce health risks for logging personnel and to attune harvesting with the silvicultural concept.

> *Possible action* 19. To draw up detailed prescriptions, including measures such as climber cutting, marking of trees to be felled and/or residuals to be retained and indications of extraction direction and felling direction.

3.2.2 Roads

Principle 23. Planning, location, design, and construction of roads, bridges, causeways and fords should be done so as to minimize environmental damage.

> *Possible action* 20. Limits to dimension, road grades, drainage requirements and conservation of buffer strips along streams should be specified (see further Appendix 3).

3.2.3 Extraction

Principle 24. Extraction frequently involves the use of heavy machinery and, therefore, precautions must be taken to avoid damage.

> *Possible action* 21. A logging plan should be drawn up including:
> - areas where logging is subject to special restrictions or forbidden (flora and fauna conservation and soil protection areas, buffer strips, sites of cultural interest);
> - specifications for construction and restoration of skidding tracks, watercourse crossings and log landing (including drainage);
> - wet weather limitations;
> - allowed harvesting equipment;
> - machine operator responsibilities (directional felling, etc.); marking of trees to be retained and trees to be removed (see further Appendix 3).

3.2.4 Post-harvest stand management

Principle 25. Post-harvest operations are necessary to assess logging damage, the state of forest regeneration, the need for releasing and other silvicultural operations to assure the future timber crop.

> *Possible action* 22. Carry our post-harvest inventory, establishing the need for silvicultural interventions.

3.3 PROTECTION

3.3.1 Control of access

Principle 26. Permanent production forest should be protected from activities that are incompatible with sustainable timber production, such as the encroachment by shifting cultivators often associated with the opening up of the forest.

> *Possible action* 23. Access to logging roads that are not part of the national infrastructure (i.e. through-roads) should be strictly controlled. Consideration should be given to the possibility of managing special buffer zones bordering the production forest for the benefit of the local population.

3.3.2 Fire

Principle 27. Fire is a serious threat to future productivity and environmental quality of the forest. Increased fire risk in areas being logged, and even more so in areas which have been logged, demands stringent safety measures.

> *Possible action* 24. A fire management plan should be established for each forest management unit, taking into account the degree of risks. The fire management plans may include regular clearing of boundaries between the forest estate and other areas, and between forest blocks within the forest estate. In areas being logged or already logged, additional safety measures such as restrictions on use of fire, keeping corridors between blocks free of logging debris, etc., should be specified. Advance warning systems, including those that are satellite based, should be used.

3.3.3 Chemicals

Principle 28. Chemicals, such as the ones used in silvicultural treatment, constitute risks both in terms of personnel safety and environmental pollution.

> *Possible action* 25. Instructions for handling and storage of chemicals and waste oil should be provided and enforced. Special restrictions are to apply near watercourses and other sensitive areas.

3.4 LEGAL ARRANGEMENTS

3.4.1 Concession agreements

Principle 29. There should be incentives to support long-term sustainable forest management for all parties involved. Concessionaires should have the long-term viability of their concession provided for (mainly by government controlling access to the forest); local population should benefit from forest management (see section 4); government should receive sufficient revenue to continue its forest management operations.

> *Possible action* 26. Concession legislation should be adopted or reinforced to cover the following aspects: the responsibilities and authority of the forest service and the responsibility of the concessionaires; the size and duration of concession or licence; conditions for renewal and termination.
> Concession legislation is to include (see also Appendix 4): (a) categories of contracts, and application and granting procedures; (b) objects of the contract; (c) rights granted and rights withheld; (d) establishment or expansion of local wood-processing units; (e) felling, wood extraction and transport; (f) road construction and improvement of infrastructure; (g) forest management and reforestation; (h) forest taxes, stumpage and other fees; (i) control, supervision, and sanctions for disrespect of concession terms; (j) other general provisions; (k) other environmental considerations.

3.4.2 Logging permits on private or customarily held land

Principle 30. For private or customarily held forests the basic approach to sustainability is the same as for government forests (see 3.4.1).

Principle 31. The national forest service should provide assistance to customary rights holders and private forest owners to manage the forests sustainably.

> *Possible action* 27. Provide for or strengthen a forestry extension service which can provide forest management training for various categories of land-holders.

3.4.3 Salvage permits

Principle 32. Timber from forest land to be converted to other uses, and from forest damaged by hurricanes and other disasters, should be optimally utilized. At the same time, disruption of management of the permanent production forest should be prevented.

> *Possible action* 28. Devise mechanisms to provide for orderly introduction of timber from salvage operation into the market.

> *Possible action* 29. Provide for volume adjustment of log removal from logging concession to account for timber, including material of below-minimum exploitable diameter, becoming available from conversion land.

3.5 MONITORING AND RESEARCH

Principle 33. Monitoring and research should provide feedback about the compatibility of forest management operations with the objectives of sustainable timber production and other forest uses.

3.5.1 Yield control and silviculture

> *Possible action* 30. Develop design of Permanent Sample Plot (PSP) procedure (distribution, number, design, minimum measurements) and of monitoring of PSPs to increase accuracy of Annual Allowable Cut calculations.

> *Possible action* 31. Assessment of compatibility of management practices and silvicultural systems by carrying out regeneration surveys, and studies on need for post-harvest stand treatment and other relevant subjects.

> *Possible action* 32. To study the dynamics of main timber species to enable stand modelling.

3.5.2 Environmental impact studies

> *Possible action* 33. To assess compatibility of logging practices with declared secondary objectives such as conservation and protection, and with the overall principle of sustainability.

4 Socio-Economic and Financial Aspects

Principle 34. Sustained timber production depends on an equitable distribution of incentives, costs and benefits, associated with forest management, between the principal participants, namely the forest authority, forest owners, concessionaires and local communities.

4.1 RELATIONS WITH LOCAL POPULATIONS

Principle 35. The success of forest management for sustained timber production depends to a considerable degree on its compatibility with the interests of local populations.

Principle 36. Timber permits for areas inhabited by indigenous peoples should take into consideration the conditions recommended by the World Bank and the ILO for work in such areas inter alia.

Possible action 34. Provisions should be made: for consultation with local people, starting in the planning phase before road building and logging commences; for continued exercise of customary rights; for concession agreements and other logging permits to cover the extent of assistance, employment, compensation, etc., to be provided.

4.2 ECONOMICS, INCENTIVES, TAXATION

Principle 37. Management for timber production can only be sustained in the long term if it is economically viable, (taking full account in the economic value of all relevant costs and benefits from the conservation of the forest and its ecological and environmental influences).

Possible action 35. National and international marketing efforts should be intensified in order to realize highest possible value of forest products and improve utilization of the resources from sustainably managed forests.

Principle 38. A share of the financial benefits accruing from timber harvesting should be considered and used as funds for maintaining the productive capacity of the forest resource.

Principle 39. Forest fees and taxes should be considered as incentives to encourage more rational and less wasteful forest utilization and the establishment of an efficient processing industry, and to discourage high-grading and logging of forests which are marginal for timber production. They should be and remain directly related to the real cost of forest management. Taxation procedures should be as simple as possible and clear to all parties involved.

Principle 40. In order to achieve the main principle of good and sustainable management, forest fees and taxes may need to be revised at relatively short notice, due to circumstances outside the control of loggers and the forest agency (e.g. fluctuations in international timber market and currency). The national forest agency should be granted the authority to carry out such revisions.

Principle 41. Continuity of operations is essential for sustainable forest management.

Possible action 36. In order to remain operational even in adverse budget situations, the forest authority should be granted a certain degree of financial autonomy which, among other things, should allow the accumulation of funds. This can be achieved e.g. by allowing the forest authority to collect part and maybe the full amount of forest fees and taxes without intervention from other government departments.

Appendix 1.3
ITTO Guidelines for the Establishment and Sustainable Management of Planted Tropical Forests

ITTO POLICY DEVELOPMENT SERIES 4

ITTO, 1993

CONTENTS

1 Introduction

Planted forests are an important element of land use in the tropical world. Planted forests can fulfill many of the productive and protective roles of the natural forest. When they are adequately planned, planted forests can help stabilize and improve the environment. However, conservation of local plant and animal species and ecosystems and ensuring ecological stability at the landscape level require complementary action within integrated land-use and development plans.

Deforestation of all kinds is increasingly becoming a major problem in the world. The world's population is anticipated to double in the next 60 years and social and economic development will increase demand for and consumption of wood products. This demand can only be met by appropriate forest conservation and development, including the establishment and improved silvicultural management of plantations.

Combatting desertification and soil erosion may also require the establishment of protective and productive forest plantation. The conservation and preservation of natural forest will certainly become more difficult if complementary plantation forests are not established in an appropriate manner and scale. On the other hand, it would be wrong to assume that planted forests could substitute for natural forests and replace them as a source of raw materials and environmental and social benefits. Such assumptions could lead to natural tropical forests being cleared to provide sites for industrial forest plantations which promise to produce much higher volumes of timber per unit area. However, major social conflicts may also arise from industrial plantations displacing existing landholders and disrupting prevailing patterns of land use. Possible detrimental environmental and ecological effects of large-scale introductions of exotic tree species are also emerging as major concerns and policy issues in some tropical countries and amongst the international community.

Planted tropical forests can achieve extremely high levels of timber production and may therefore offer tropical countries a considerable competitive advantage in the international timber trade. However, despite rapid initial growth, many tropical plantation forests have not fulfilled their early promise and a number of significant problems have arisen.

Some tropical plantations have caused environmental problems through reductions in local biodiversity at both the level of plant and animal species and the landscape level. Poorly designed plantations may even accelerate erosion, water pollution and stream-bed sedimentation. In other cases, plantations have been planted but not adequately maintained. In still other cases, plantation forests have successfully reached maturity only to find that there were no markets for the species that were grown. There is therefore a real need to ensure that the establishment of industrial tropical timber plantations does not lead to the over-production of particular species or classes of forest products similar to the over-production which has occurred with many agricultural plantation crops in the tropics with such devastating economic consequences.

These Guidelines on sustainable management of planted tropical forests have been prepared to help promote sustainability in all aspects of tropical forest management and to help solve existing problems. They have also been prepared to help prevent repeating mistakes made earlier elsewhere. It is hoped that the production of the Guidelines may:

- stimulate policy development and the adoption of comprehensive planning processes;
- help to ensure environmentally and socially acceptable selection of site, species and forest design;
- help to adopt appropriate procedures of establishment and management for all types of planted forests in the tropics;
- help planners to reduce the risk of selecting unsuitable species, provenances or populations (clones);
- stimulate the adoption of appropriate management throughout the whole of the life of the planted forest with particular emphasis on the often neglected post-establishment period;
- focus the attention of forest managers and planners on the importance of pre-establishment and continued market evaluation and the ultimate end use of the forest products they are attempting to grow;
- help to prevent the misallocation of scarce human, land and financial resources.

These Guidelines present fundamental concepts, expressed as a set of principles and recommended actions. The Guidelines constitute the international reference standard established by ITTO for the development of more specific guidelines, at the national level, for the sustainable establishment and management of planted tropical forests for timber production and other purposes. The development, application, adherence and enforcement of national guidelines based on this standard are matters for national decision-making by individual timber producing countries.

Tree planting activities encompass diverse agents: state administration and public agencies, large or small industrial and commercial enterprises, regional authorities, communities and individuals. While the guidelines present universally relevant fundamental principles, their specific application depends on the particular condition of site and case with respect to the natural, technological, economic, societal-cultural, and socio-political situation. The eventual role of a planted forest in the general pattern of resource use depends on a mix of social, economic and environmental factors. Decisions on location, site, species, silviculture, management and objectives must therefore comply with local and national political, social, economic and environmental conditions. Of central importance are the purpose and functions of the planted forest and the way by which these are achieved.

ITTO attaches high priority to the definition of the essential principles and associated actions which should serve to guide the development of national guidelines in each country so that they may conform to the international reference standard agreed within the Organization. The Organization also gives high priority to assisting member countries, which may need and request assistance, to obtain such outside technical and financial help as they may require to develop their own national guidelines.

The ITTO Guidelines are presented in the form of general principles and recommended actions ranging from general policy considerations to aspects of operational forestry. They are relevant to any deliberate planting of trees in tropical environments. However, they outline principles and actions that should be particularly relevant to the establishment of intensively managed large-scale plantation forests for industrial wood production.

2 Policy and Legislation

2.1 FOREST POLICY

Principle 1 The forest sector offers major opportunities for sustainable socio-economic development and the improvement of the quality of life in tropical countries. All countries therefore need to understand both the existing and future demands for all benefits, goods and services from all types and categories of forests. Governments and people must clearly judge and understand the capacity of their forests and forest lands to provide these benefits, goods and services. In particular, studies of both the domestic and export sectors are required to define the country's need for planted forests to complement and supplement their natural forests in all respects, including as a resource base in a long-term wood production strategy.

Natural forests provide optimal comprehensive environmental and habitat protection. A first priority should be to maintain and restore natural vegetation cover. Even when natural forest cover is degraded, rehabilitation by natural regeneration rather than replanting is generally the preferred option. Compatible forms of timber harvesting and production management on suitable sites could play a positive role in forest areas where the main management objective is environmental protection.

> *Recommended action* 1 Undertake comprehensive studies to determine:
> - the demand for forest products in both the domestic and export sectors;
> - the demand for environmental protection;
> - the capacity of the existing forest estate to provide these goods and services;
> - the location and extent of the planted forest estate that will be needed to supplement existing forests to meet these production targets and community demands for environmental services in a sustainable manner;
> - the dependence and demands of local communities for economic, spiritual and cultural values on the types of forest lands under consideration for reforestation.

Principle 2 Provisions for the establishment and sustainable management of planted forests must be considered in the context of an integrated land use plan for national economic and social development. Thus, planted forests should normally only be established on lands which are known to be capable of supporting all aspects of their long-term management and utilization without land degradation. The creation of plantations must be balanced with the needs for protection of the site and the environment, the conservation of biological diversity of all types, the needs and aspirations of the present people and the potential demands of future generations.

In particular, any larger-scale plantation scheme must incorporate provisions to meet the needs for site conservation and protection of the environment, customary and statutory land rights and the subsistence needs of local communities.

Principle 3 Strong and continued political commitment at all levels is indispensable for the successful establishment and management of planted forests at the national, management unit and local levels.

Principle 4 Institutional capacities must be established to allow for the development and implementation of the integrated land use plans necessary for effective forest establishment and management.

Principle 5 Effective community consultation procedures are an essential component of these institutional planning processes.

Recommended action 2 Formulate and implement a national land use policy which promotes the sustainable use of all natural resources, including the establishment of a permanent forest estate. Ensure that national land use policies are in turn an integral part of national socio-economic strategies and plans for the development of industry and employment.

Recommended action 3 Formulate and implement a national forest policy as an integral part of the national land use policy to ensure a balanced use of forest resources. Develop national forest policies through systematic consultation with and the consensus of all parties involved or affected, including: national government agencies, provincial and local government, local communities, non-governmental organizations, scientific experts, and the private business sector.

Recommended action 4 Conduct seminars to discuss policies about land use, land allocations and the role of planted forests in the national economic and social environment. These seminars should involve all the above-mentioned sectors and interest groups.

2.2 LEGISLATION

Principle 6 The national forest policy must be supported by appropriate legislation which, in turn, should be in harmony with laws concerning related sectors. Sufficient resources must also be allocated on a continuing basis to ensure that legislation and policies are effectively implemented.

Recommended action 5 Enact laws and regulations at appropriate national and sub-national government levels to support the established forest policy, in harmony with the policies, laws and regulations of related sectors. Achieving such harmonization will often involve the repeal or revision of existing laws and regulations, both inside and outside of the forestry sector.

Recommended action 6 Repeal or revise existing policies and laws that provide incentives for the wasteful use and inappropriate degradation of forested lands. In addition, revise all laws and government policies to both encourage long-term investment in forests and forest-based industries and remove any disincentives to such investments. Provide government assistance to help investors meet wider environmental and socio-economic obligations, both through the provision of financial incentives and the development of appropriate conditions for investment security.

Principle 7 There must be mechanisms for the regular revision of policy and legislation in the light of new social, economic and environmental circumstances and/or the availability of new information.

Recommended action 7 Provide adequate funds for research, monitoring and continuing community consultation to allow informed updating of policies and practices. In particular, provide for the design and implementation of tariff and tax systems, especially for export forest products. Such systems should allow for sufficient adjustment to changes in the interdependent international and domestic market situations, and consider site disadvantage, such as nature and distances of access.

2.3 NATIONAL FOREST INVENTORY IN RELATION TO LAND ASSESSMENT SURVEYS

Principle 8 A national forest growing stock and land inventory should establish the status of all forests, independent of ownership status. Land tenure, the land development plans of other agencies and customary rights will frequently be a key item in forest land inventory data in many tropical countries. The national forest inventory should provide a clear picture of the legal and ecological status of forests under various forms of land tenure and customary rights. The results of such inventories should be evaluated with the results of broader land use studies to determine the potential opportunities for and constraints on the development of planted forests. Forest land and growing stock inventories must apply techniques which ensure reliability, continuity, accuracy and sufficiency of data.

Principle 9 There should be flexible provisions for such inventories to be broadened to include information not previously covered, if and when the need and opportunity for the collection of such additional information arises.

2.4 PERMANENT FOREST ESTATE

Principle 10 Certain categories of land, whether public or private, need to be kept under permanent forest cover to secure their optimal contribution to national development and environmental protection – see Appendix 1.

Principle 11 Land allocation for the establishment of planted forests must consider the interests, legal rights and long-term plans of all sectors concerned with or affected by their development. Particular attention must be given to the interests of the local residents and communities who will experience most closely any changes brought about by particular planted forest proposals. There will therefore be a need for specific planning activities at the national, regional and local levels.

> *Recommended action* 8 Identify, survey and delineate the various categories of sites, and allocate land to the various forms of forest in consultation with affected communities giving careful consideration to their legal claims on the land. In these evaluations, take into account present and future needs for agricultural and pasture lands, as well as customary use of various forest products and conservation requirements. Define the role of planted forests in achieving or supporting this optimal pattern of forest land allocation needs.

Principle 12 Natural forest should not be cleared for the establishment of planted forests unless this is proved to be essential to justify retaining the land under forest cover.
   The feasibility, desirability and necessity of replacing existing natural or secondary forest by planted forest should be expertly assessed in a manner that ensures independence of judgement. This assessment should include the full range of ecological, environmental, economic and social implications and their long-term consequences. The advantages and disadvantages of natural forest regrowth, enriched natural regrowth and planted forests respectively, should always be comprehensively compared before forest allocation and management decisions are taken. These issues should also be subject to wide community consultation and discussion to ensure that forest management decisions meet community needs and are socially acceptable – key pre-requisites for sustainable forest establishment and management.

Principle 13 Notwithstanding the provisions of Principle 12, where natural forest areas have been so severely degraded by past land use practices that their effective recovery and survival as forests is in doubt, consideration should be given to conversion of suitable sections of these highly degraded areas to more productive planted forests.
   In cases where careful assessment has proved that conversion of severely degraded natural or secondary forests to planted forests is justified, fully considering environmental, economic and social factors, such highly degraded areas may be transformed to forests or to suitable forms of agroforestry systems with improved and sustainably high levels of productivity. This would contribute to the overall forest and agricultural productivity and thereby can reduce pressures on remaining natural forests. The overall benefits the community and the nation gains from the forest estate would thereby increase in the long run.

Recommended action 9 Continuously monitor the condition of the whole forest estate and revise forest management plans in consultation with the community to promote the efficient and well-balanced use of forest lands.

2.5 LAND OWNERSHIP

Principle 14 The principles and recommendations outlined in these guidelines for planted forests should be equally applied to publicly and privately owned lands and to lands controlled by customary rights.

Principle 15 The principles and recommended actions outlined in these guidelines can only be implemented if there is secure and long-term tenure of the land, and acceptance of suggested land use changes and land allocations by the local communities. In particular, claims based on legal titles or on statutory or customary rights in all types of forest lands must be duly considered, including claims to ancestral territories and cultural sites. These and other local claims recognized under national law also require consideration in relation to environmental protection, sustainable economic development and compensation for being displaced or otherwise impaired by the establishment of planted forest areas:

> *Recommended action* 10 Develop comprehensive land allocation plans, legal instruments and investment incentives to protect any permanent forest land tenures. Create and maintain appropriate local institutions to both monitor the implementation of land allocation plans and enforce any legal instruments that may be required. Involve affected people in the development and implementation of such land allocation plans and subject these plans to wide community discussion and review before their approval and implementation.

2.6 NATIONAL FOREST SERVICE

Principle 16 There should be a national agency capable of effective, integrated management of the public forest estate. Such an agency must also be capable of assisting in the establishment and management of all types of forest on communal and private lands, according to the objectives and principles laid down in the national forest policy.

> *Recommended action* 11 Provide for a national agency to carry out the above functions.

> *Recommended action* 12 Provide the staff of such an agency with appropriate training and adequate resources to allow them to carry out their duties effectively and efficiently. Where communal or private forest lands are likely to be involved, special emphasis should be placed on developing extension systems and the community communication skills necessary for effective participatory planning.

3 Feasibility Assessment

3.1 ENVIRONMENTAL CONSIDERATIONS

Principle 17 Planting trees will usually induce changes in the local biological and physical environment. These changes can be potentially beneficial or harmful, or both.

> *Recommended action* 13 Include comprehensive environmental impact assessment procedures in all pre-planting feasibility investigations. Promote the positive impacts of change while simultaneously minimizing any adverse impacts, so as to increase the overall benefits of the proposed planted forest to the community.

Principle 18 In many environmentally important areas such as steep slopes, catchment areas and degraded watersheds, the establishment of a well-managed forest cover offers many environmental, social and economic advantages. Similarly, well-designed and well-managed planted forests can provide appropriate protection and help to stabilize and restore fragile and degraded areas.

Recommended action 14 Assess the feasibility of establishing planted forest programmes on such lands with particularly important environmental functions, giving recognition to these potential advantages. Such assessments should also recognize that possible environmental restrictions may need to be placed on future harvesting practices, ranging from restrictions on the actual areas eventually harvested, to restrictions on the type of machinery that can be used on particularly sensitive sites. In conducting feasibility studies for these areas, determine whether or not the costs of such likely restrictions on utilization are counter-balanced by the direct and indirect benefits afforded by the protective functions and positive environmental effects of the planted forest.

Principle 19 Replacement of natural vegetation by planted forest can simplify existing ecosystems. Although plantations can be designed to contribute to the conservation and enhancement of genetic resources of given target species, their possible negative impacts on ecosystem conservation and overall biodiversity must be carefully assessed. The risks of deterioration of biodiversity should be reduced by appropriate silvicultural management. Important management considerations include the appropriate siting of planted forests and setting aside other land areas primarily for the conservation of regional biodiversity. It should be remembered that planted forests which are designed to provide a diverse habitat, such as mixed, multi-storied forests, are richer in animal and plant species which regulate them whereas uniform, simple monocultural designs create forests which require constant human inputs to maintain viability and productivity.

Recommended action 15 In evaluating the feasibility of particular proposals to establish planted forests, carefully consider the impact of land use allocations, the actual siting of the forest land and its detailed compartment design on the local and regional patterns of species distribution and biodiversity and on the local and regional climate (micro- and the meso-climate).

Principle 20 Potential planting sites can have attributes that have archeological, cultural or spiritual significance at the local, national and global levels that may be adversely affected by forestation activities.

Recommended action 16 Include the identification, description and evaluation of the significance of such sites in all pre-planting resource inventories. Consider appropriately the resource values to all concerned parties in planning land allocation and establishment of forest plantations.

Principle 21 Natural and planted forests store carbon and exchange a multitude of trace gasses with the atmosphere. They thereby affect micro- and meso-climate and, to a smaller measure, macro-climate. The on-going climate change and warming of the surface water of the tropical oceans will make the tropical climate more variable and extreme events more severe. Forest trees and whole forest eco-systems may therefore be expected to suffer increasingly from stress and damage. Precautionary adaptation appears prudent and should be considered in site selection and crop designing.

Recommended action 17 Give consideration to these issues in general land use planning when determining both regional and national planting goals and when specific decisions are taken on forest structure and silvicultural and agroforestry systems for the various types of sites. The objective should be a precautionary adaptation of the forests to expected future environmental conditions such as more violent storms and floods, and more severe droughts. Questions on the preservation of suitable germplasm in-situ or elsewhere are relevant in this respect.

3.2 SOCIO-ECONOMIC CONSIDERATIONS

Principle 22 Planting trees can decisively affect social and economic conditions at the national as well as regional and local levels. These effects can be either positive or negative.

Positive effects can range from enhancement of social and economic development through provision of local access to resources; employment generation and the creation of investment opportunities; increased potential for industrial development; the possibility of increasing and stabilizing export earnings; and the subsequent improvement of rural life through better access to enhanced infrastructures, educational opportunities and medical care.

Negative socio-economic impacts can range from disruption of traditional land rights and patterns of land use, reduction of cultural values, inefficient use of investment funds through the development of forest resources not sufficiently targeted to market demand, and possible extra-regional impacts through economic displacement of competing forest enterprises. Local communities experience the most direct beneficial or detrimental impacts. Their views and needs should receive particular attention because their acceptance and cooperation is essential for success.

> *Recommended action* 18 Include comprehensive social and economic impact assessment procedures in all pre-planting feasibility investigations. Promote the positive impacts of change while simultaneously minimizing any adverse impacts, so as to increase the overall benefits of the proposed planted forest to the community.

> *Recommended action* 19 Diversify crop types and their site-specific location in the area to meet community demands. Provide for intercropping (taungya, tumpangsari) and mixed cropping (agroforestry) on suitable areas within a plantation scheme to provide ecological and economic benefits that will enhance acceptability to the local community and reduce costs for silvicultural management and protection. Incorporate fuel, fruit or fodder trees in the forest stand for local usage, and allow the supply of timber for domestic use to benefit the community and in return benefit plantation management.

> *Recommended action* 20 Include detailed market evaluations in all planted forest feasibility studies concerning planted forests of any type, including single-species plantations, mixed forest and agroforestry plantations.

3.3 INSTITUTIONAL CONSIDERATIONS

Principle 23 To succeed at the social, technical and economic levels, and to be environmentally sustainable, planted forest programmes must be supported by strong national and local institutions to ensure integrated planning, community involvement and monitoring of the economic and technical feasibility and performance of all management activities.

The strength of these institutions depends on the strength of the political support they receive. Government institutions must receive adequate financial support, adequate staff, stable employment and attractive career opportunities. The staff should be structured to cover all necessary fields of development, research and extension.

Principle 24 Efficiency must be maintained by adequate in-service and special training. Continuous interchange of information and experiences within and between national institutions and with foreign institutions is necessary to maintain expertise at sufficiently high levels.

Principle 25 Non-governmental organizations (NGOs) can play important roles in forest development programmes as partners of government and local communities in feasibility studies and the planning process, as sources of information and often as innovating elements. Their participation should provide information and other resources which constructively contribute to balanced land use planning.

Recommended action 21 Implement institutional strengthening and promote participative procedures at all levels, but especially at the local level, to improve efficiency and expertise.

4 Planted Forest Establishment

4.1 MANAGEMENT PLAN PREPARATION

4.1.1 The importance of management planning

Principle 26 Integrated planning at all levels reduces private and public economic and environmental costs. A management plan is therefore an essential component of the establishment and sustainable management of any planted forest, and must complement other relevant plans in related sectors.

> *Recommended action* 22 Make provision for adequate management planning at all levels of forest management.

> *Recommended action* 23 Forest management plans should address at least the following topics:
> - Areas to be excluded from planting and production management, including steep topography, fragile soils, protective beds along watercourses, areas for the preservation of amenity and areas for nature, species and genotype conservation.
> - The layout of the road, fire protection and extraction network.
> - Procedures for site preparation; planting; tending; prevention of erosion, compaction and other forms of site degradation; silvicultural treatments; and controlled burning.
> - Fire protection and fire management.
> - Biological pest management and protection against pests, diseases and climatic calamities.
> - Market development and utilization plan.
> - Provision of all kinds of forest benefits to the local communities and recognition of customary rights.

Principle 27 Management of planted forests should implement traditional multiple-use principles in order to produce multiple benefits. Management objectives should therefore consider all forest values determined by comprehensive evaluation. Objectives for particular areas will have to be specific to the site and environmental, economic and social circumstances. The objectives must relate appropriately to the goals of the plantation programme and the interests of the local communities. During planning and establishment, these communities should be actively involved in order to motivate them to continued cooperation and eventually increase their income and quality of life. Such involvement could take the form of allowing local people to cultivate agricultural crops between tree plantation for a certain number of years and of continued employment in forest operations.

> *Recommended action* 24 Prepare or update existing 'codes of best practice for forest conservation and management' to describe the procedures for planning and monitoring for all planted forest in order to ensure completeness of prescriptions and integration with other forms of land use.

4.1.2 Soil and site considerations

Principle 28 In general, the better the soil and site conditions are, the lower the risk will be of land degradation from all forms of land use. This generalization is particularly applicable to intensive cultivation of single-species plantation forests. Unfavourable sites, such as steep slopes, fragile or deficient soils, carry high rates of risk and low rates of productivity and should be reserved for protection and conservation forestry.

Recommended action 25 Carry out well-designed site mapping in order to carefully determine the suitability of the site based on adequate site and soil classification and survey procedures. Give particular consideration to both risk and production potential in the site-specific allocations of different forest crops to various types of soil and site. Restrict intensive forestry, especially short-rotation, single-species industrial timber plantations to sites with physically, chemically and biologically favourable soils and flat to gently sloping terrain.

Principle 29 The establishment of a productive forest plantation on degraded land usually requires either a prior phase of a sown or planted pioneer vegetative or a usually expensive artificial amelioration of soil fertility. After afforestation, repeated biological or artificial replenishment of nutrient stock may be necessary in the course of time, to prevent impoverishment of the soil through nutrient losses from leaching, erosion and harvesting. The natural process of restoring soil fertility by secondary forest growth during well-managed fallow periods such as in the various forms of the taungya (agrosilviculture) systems and the traditional systems of shifting swidden cultivation can serve as a model for sustainable silvicultural methods. The degradation of soil and vegetation in short-rotation swidden also demonstrates the consequences of overexploiting the site potential.

Recommended action 26 Assess the biological soil activity and the nutrient status of soil before afforestation with the objective of designing an adjusted soil amelioration scheme. Regularly monitor the status of the soil and the health of the growing stock.

Principle 30 The activity of soil fauna, flora and microbes is an essential element of soil fertility which requires careful maintenance. The maintenance of adequate conditions of soil biology is a key element of sustainability in humid tropical forest management.

Recommended action 27 Minimize soil exposure at the time of initial afforestation and through subsequent forest management activities.

Recommended action 28 Maintain an effective soil cover to reduce erosion and supply adequate amounts of appropriately mixed organic material to the soil, either by developing and maintaining a diverse, layered forest structure, or by soil-cover crops or by intercropping in single-species uniform industrial plantations.

Principle 31 Together with litter supply and humus formation, the interaction of soil porosity and texture with other physical conditions is a most important factor affecting the effective fertility and biological activity of tropical soils. The use of appropriate technology and careful planning is essential to maintain an appropriate soil environment for sustainable forest management.

Recommended action 29 Avoid soil erosion and compaction from inappropriate establishment procedures such as the use of excessively heavy machinery or the use of intensive cultivation practices on land not suited to this purpose. Undertake land capability and land suitability appraisals as a key component of planning for the planted forest, to ensure that intensive cultivation and other site preparation practices are restricted to sites capable of supporting these activities without causing land degradation.

Principle 32 Non-crop species of trees and shrubs may have important ecological functions such as fostering the development of symbiotic relationships, improving soil cover and litter diversity and providing habitat for other members of the trophic network. As such, they should not automatically be considered as weeds that have to be immediately eliminated. Rather, their potential benefits should be carefully balanced against the cost of possible competition effects, so that investments in weed control can be focused onto crucial stages of the forest's development. This will help to ensure that funds are used efficiently and tending does not become counter-productive.

Recommended action 30 Wherever feasible, manage and integrate spontaneous auxiliary vegetation into silvicultural practice rather than automatically eliminating such vegetation. Introduce auxiliary vegetation where ecological stability reasons require such actions.

Recommended action 31 Control vegetation which is detrimental from an economical and managerial point of view by low impact methods such as, controlled burning or manual slashing. Minimize the use of chemicals or intensive mechanical disturbance and compaction of the soil, as they may have undesirable effects such as inducing accelerated surface water run-off and erosion or the pollution of adjacent water courses. Avoid the use of chemicals with residual toxic properties and take particular care when their use becomes necessary.

4.1.3 Research needs

Principle 33 Basic scientific and application-oriented research is the fundamental source of the information needed for sustainable timber production and other forms of forest use. The performance of the forest crop, the impact and effectiveness of forest management operations and the status of soil and site conditions all need to be continuously monitored so that timely corrective measures can be taken in response to any long-term trends of change. Research is also needed to monitor changes in community needs and expectations at the local, regional and national levels together with other aspects of the socio-economic environment in which forest managers must operate.

Recommended action 32 Where possible and feasible, take advantage of developments of sophisticated methodologies such as systems analysis, ecological modelling, ecosystem level studies and information systems to improve the information base for decision-making in forest management. Ensure that resource economics and the social sciences are an integral part of any research programme associated with planted forests.

4.2 TECHNICAL REQUIREMENTS

4.2.1 Choice of site, tree species and planting material

Principle 34 The selection of sites must fully and comprehensively consider natural site conditions, logistic and economic features of the site as well as the social and political environment. In principle, production forests should be as close to the existing markets as other competing land users permit. Once the site is determined, the selection of species and genotype must be carefully matched with the site conditions. Besides considering soil type and average climatic conditions, consideration must also be given to ecological risk factors and extreme events such as excessive rainfall, floods, droughts, cyclonic or convective storms and biotic hazards. These site conditions and risk factors, and competing vegetation should influence the choice of the most suitable type of planting material. In any case, the origin of the seed, plants or cuttings must be identified, certified and labelled. Plantations must only be established with source-identified genetic material.

Recommended action 33 Select a suitable soil classification and undertake a careful and comprehensive soil and site survey. This is needed to provide the essential basis for the choice of species, species allocation to the site units, and the choice of adapted designs of the structure of the forest crop in order to achieve the highest possible productivity at acceptable low levels of risk. Following these basic decisions, evaluate the range of available planting materials and methods, such as direct seeding, bare rooted seedlings, potted or tubed materials, cuttings of different sizes, provenances and clonal material, and select the most suitable site-specific combination.

Principle 35 Exotic species often have initial superior performance and management advantages as they can grow in the absence of a suite of locally adapted crop predators. However, through the processes of natural selection, domestic tree and shrub species may have developed adaptations to the local edaphic, climatic and biotic conditions which may give them long-term advantages over exotics. Proof of such long-term adaption is usually either lacking or restricted to a few, short-term examples for exotic species in most areas. Forest managers and planners should therefore not assume that the initial growth advantages of exotic species will be able to be maintained without additional management inputs over time.

In the comparative assessment of the potential of native species, it must be considered that when these species are planted in the open and grown in single-species uniform plantations, they often may behave and perform quite differently than in their original habitat. It must also be remembered that practical experience with native species in the past usually has been gained under specific and often unique circumstances of site and conditions for growth. Such experience therefore can be very misleading. Colloquial experience cannot replace the systematic knowledge gained from well-designed scientific research.

Considerable improvement of growth, yield, produced quality, adaptability to sites, and resistance to pests and diseases are possible through tree breeding. Such improvements are especially important to industrial plantations when economic returns can be demonstrated to outweigh the additional cost of genetically improved material.

> *Recommended action* 34 Give preference to the use of native tree species wherever feasible and practical both for production of high volumes and for the production of quality timber and high value. In this regard, the ease of obtaining seed and propagating commonly planted exotic timber species such as pines, gmelinas, eucalypts, acacias and teak should be compared with the benefits of the likely long-term adaption to the site by local species.

> *Recommended action* 35 Carry out research on promising native species in order to establish their real potential as planted forest species for high quality timber production. In particular, research on genetic improvement should be initiated or expanded. This applies also to the genetic improvement of species for agroforestry systems and for environmental improvement.

> *Recommended action* 36 When exotic species are chosen, care must be taken to reduce risks by:
> * allocation to suitable soil/site units;
> * careful screening of provenances, hybrid and clone properties in relation to site requirements, growth performance and product quality;
> * mixing species, provenances or genotypes (clones) both spatially within the same contemporary crop area and temporally with mixed sequences of crops for the site in a rotational silvicultural system.

4.2.2 Roads and site protection

Principle 36 The planning, location, design, and construction of roads, bridge structures and other forest management infrastructure should be done so as to minimize erosion and other forms of damage to the site and the wider environment, and allow easy access for all forest operations, including fire management. However, roads constructed to facilitate access and management of planted forests may provide undesirable access to adjacent areas of the natural forests.

> *Recommended action* 37 Prepare prescriptions for road design, drainage, fire protection and other infrastructure requirements appropriate for local conditions. Provide for prevention of the undesirable use of roads to access natural forest, wildlife or nature protection areas.

Principle 37 Sufficiently broad, completely protected buffer strips along stream banks and adjacent riparian areas should be kept under special management to reduce sediment and nutrient inflow into adjacent water courses.

> *Recommended action* 38 Prevent soil disturbance and maintain an appropriate width of undisturbed vegetation along all water courses and riparian zones to maximize the absorption of overland water flow, nutrients and sediment from disturbed sites in adjacent production forest areas.

Principle 38 Fire can be a serious threat to the productivity, ecological stability and social and environmental quality of planted forests and their growing stock. Fire risks may increase as both living and dead biomass accumulates during the course of planted forest's development. In some areas; fire risk may also increase during the life of a single rotation of the planted forest in response to climate change associated with global warming. Fire risks and fire management requirements will also generally increase with the size of the planted forest estate.

Principle 39 There is a growing body of knowledge about both the significance and behaviour of fire in forest ecosystems. This knowledge provides a basis for minimizing fire risks and planning fire management strategies.

> *Recommended action* 39 Prepare a fire management and fire suppression plan for each planted forest area taking into account the value of the planted forest estate and the degree of risk associated with local conditions. Such fire management plans should at least allow for:
> - the establishment and maintenance of cleared fire breaks along boundaries between the forest estate and other areas, and between forest management units within the forest estate;
> - the establishment and maintenance of fire resistant species or vegetation types within or between planted forest units;
> - a collection and retrieval system for fire weather and fuel data to allow for efficient fire danger and fire behaviour prediction;
> - a fire detection and warning system;
> - a communication strategy for forest users and adjacent residents for periods when restrictions on public access or behaviour are required due to either high fire danger or other fire management purposes;
> - a hazard reduction burning strategy;
> - a prevention and suppression strategy for wildfires including consideration of any societal causal factors;
> - a systematic fire reporting system to allow for a better understanding of forest/fire interactions in the future.

4.2.3 Site preparation

Principle 40 Appropriate site preparation can enhance the early growth and development of planted forests through amelioration and improvement of soil physical conditions and reductions in competition from other vegetation occupying the site during the establishment phase. However, the long-term effects of cultivation, drainage and other intensive forms of site preparation need to be carefully evaluated as they have a significant potential to lead to site decline and unwanted side effects.

Principle 41 Site preparation can also improve access for forest management, fire protection and eventual forest harvesting activities. It can also greatly simplify the re-establishment of subsequent forest crops in future rotations. However, poorly planned or inadequately supervised site preparation can cause serious environmental damage through soil compaction, erosion, loss of top soil nutrients and other forms of land degradation.

Recommended action 40 Undertake land capability assessments before deciding on site preparation specifications. Restrict mechanical and chemical site preparation techniques to those sites with land suitability classifications capable of supporting these activities without generating accelerated erosion or other unacceptable forms of land degradation.

Recommended action 41 Ensure that operational staff have access to all required site information and appropriate, well-maintained site preparation equipment, and that they are fully trained in recommended operating procedures. Ensure that all forest planners, managers and operators are aware of the need for soil conservation and that they are familiar with basic soil conservation principles and practices. Provide facilities for on-going training to continually improve understanding in the workforce and field practice.

4.2.4 Approach to planting

Principle 42 Planting technology is species and site specific, with the choice of technique being significantly influenced by the nature of the soil and the degree of site preparation. For example, in some cases, seedling stock may be the best choice, while in others direct sowing or the use of cuttings may be more appropriate.

Recommended action 42 Select the planting technique most appropriate to the species and local conditions. In the selection of planting technology, take into consideration factors such as soil texture, soil vulnerability and ecological fragility, topography, prevailing average climatic conditions, climatic extremes, species characteristics and the availability of labour and equipment, and the availability of finance and expertise.

4.2.5 Fertilization

Principle 43 Some nutrient supplements are usually necessary during the establishment phase to enhance growth, or in extreme cases, to ensure the survival of planted forests. This is particularly true on seriously degraded lands. Also, nutrient supplements may be necessary later in the life of a plantation, to maintain adequate diameter growth and the passing of the trees from one utilization category to the next. However, the inappropriate use of chemical fertilizers in particular can lead to a range of environmental problems ranging from the accumulation of heavy metals in soil profiles to the eutrophication of adjacent streams and waterbodies. Therefore, in designing any forest fertilization programme, priority should be given toward employing organic and biological fertilization methods. The long-term consequences of any proposed forest fertilization programme must always be carefully weighed against the short-term advantages.

Recommended action 43 Undertake land capability assessments before deciding on forest fertilization specifications. Restrict applications of chemical and organic fertilizers to sites where they are not likely to be transported to streams, waterways or groundwater as either point or diffuse non-point pollutants.

Recommended action 44 Synchronize fertilizer applications with seasonal conditions, plant growth patterns and silvicultural operations such as weed control activities that are likely to promote nutrient up-take by the growing stock of the planted forest. This will both increase the cost-effectiveness of forest fertilization programmes and reduce the risk of unwanted nutrient transport away from the application sites.

Recommended action 45 Design an integrated fertilization schedule that includes the use of biological agents such as plants and soil organisms to fix and store key elements; the use of efficient nutrient-scavenging plants; and the continuous monitoring of physical, chemical and biological soil conditions together with the nutrient status of the crop. Detailed information on integrated fertilization schedules can be obtained from the Tropical Soil Biology and Fertility Programme (see Appendix 6).

4.2.6 Tending and weed control

Principle 44 The establishment or failure to establish effective management of competing vegetation is frequently a major determinant of the success or failure of the establishment phase of planted forests in the tropics. However, as discussed in Section 4.1.2, tending prescriptions need to be developed carefully to avoid creating more problems than they solve. Effective weed control should therefore always be developed on the basis of a thorough understanding of the dynamic competition between crop species and other vegetation covers, and a thorough understanding of the short- and long-term implications of particular control strategies.

> *Recommended action* 46 Review Principles 30–32 and Recommended Actions 27–31 in Section 4.1.2, to develop appropriate practices for the management of ground vegetation and tending the tree crop. The aim should be to reduce costs and risks and improve ecological stability, environmental protective functions and habitat diversity for the particular site-specific circumstances.

4.2.7 Pest control and disease management

Principle 45 Pest control and disease management practices often become necessary to ensure the survival and effective growth of planted forests. However, many of the chemicals used for these practices can pose significant hazards to the health of both operational personnel and the wider environment through pollutant drift and reductions in local or even regional biodiversity. This in turn can lead to higher risks of new pest and disease outbreaks. Fortunately, the need to use such chemicals can be greatly reduced through the application of ecological principles in integrated pest and disease management strategies.

> *Recommended action* 47 Carefully match species, provenance and genotype (clone) selections with site conditions and cultural practices to ensure vigorous tree growth capable of resisting the pressures of pest and diseases so that the use of chemical control methods can be reduced as much as possible. Wherever possible, employ integrated pest and disease management with emphasis on biological controls. Use the potential of complex structured (layered) multi-species crops to create a diverse ecosystem which is capable of reducing pest and disease problems which are common in monocultural systems.

> *Recommended action* 48 Develop and apply forest hygiene practices that minimize the spread of fungal or insect pests and diseases.

> *Recommended action* 49 Design planted forests to enhance biological control through the provision of both floristic and structural diversity. Create obstacles for the spread of pest and disease epidemics by providing an adequate mixture and complexity of both growing stock and appropriate spatial and structural patterns of the whole forest. Recognize, however, that maintaining protective strips and reservation areas of natural forest to provide diversity within and between forest stands can also harbour predators and pests.

> *Recommended action* 50 Develop an understanding of the ecology and life cycle biology of major pest species. Combine this understanding with the principles of biological pest and disease management to avoid inadvertently favouring expansion of pest populations and to allow for more strategic applications of chemical and other control measures. Avoid indiscriminate, blanket applications of chemical control agents.

4.2.8 Staff development

Principle 46 The success of the establishment of planted forests ultimately depends on having skilled personnel at all levels of planning, management and operational activities. This requires adequate training opportunities and facilities, particularly when staff and labour are recruited locally. Working conditions, in particular safety conditions and pay scales, must be adequate and comply with internationally agreed standards.

Recommended action 51 Maintain motivation and pride of work by providing appropriate rewards, to personnel at all staff levels from forest labourers to forest managers.

5 Post-establishment Management

5.1 OPERATIONAL PLANNING

5.1.1 Preparation of work plans
Principle 47 Sustainable management is concerned with much more than just the establishment phase of the planted forest. It is concerned with management of the whole of the initial rotation and with the maintenance of site productivity for future rotations.

Principle 48 The plantation management plan should form the basis for all action and forecasts for sustainable management. It should cover at least the full initial rotation and provide a systematic framework from which the forest manager can prepare a detailed work plan. This latter document should outline the operations that have to be carried out, the resources required to undertake them and the time scale involved.

5.1.2. Institutional considerations
Principle 49 Achievement of forest management objectives requires continuity in action. As outlined for public forests in Section 2.6, there should be a national forest agency with the financial and human resources that will enable it to effectively carry out its responsibilities. For the private sector, continuity of tenure must be assured.

Principle 50 Resources should be allocated for basic and applied research programmes aimed at maximizing the efficiency of management operations and improving the productivity of the planted forest enterprise. Forestry research organizations should provide for the continuous feedback of information to forest management agencies.

Principle 51 Efficient and effective implementation of management plans and associated research programmes requires staff with high professional expertise and the ability to work with rural communities.

> *Recommended action* 52 Establish contract agreements between forest management agencies and forest research organizations for targeted research relevant to planted forest planning and management.

> Recommended action 53 Provide funding for on-going management, research and development activities in all tree planting programmes.

> *Recommended action* 54 Include high quality training and human resource development programmes as an integral part of all forest management systems.

5.1.3 Social considerations
Principle 52 The long-term success of planted forests and their management for the sustained production of timber, non-timber products and other services and benefits ultimately depends on their compatibility with the regional economy and the economic and land use policies, as well as the interest of the local and regional communities, and particularly the interests of the local people. In many cases, the effectiveness of forest management can be successfully improved by actively involving the local people and by utilizing with care and discrimination local experience.

Recommended action 55 As indicated in Section 4.1.3, keep up with the changes in the needs and aspirations of the community at diverse levels, as well as the social impact of the planted forest on the community.

5.1.4 Economic considerations

Principle 53 Management of planted forests for timber and other benefits can only be sustained in the long term if it is economically viable. Thus, monitoring the economic performance of the forest is an essential component of sustainable, research-based management. In the socio-economic context, analysis and evaluation of economic costs and benefits must include valuation of environmental services and local subsistence uses of timber and other products, wildlife and services.

Recommended action 56 Establish a system for continuous and comprehensive collection of data on all costs and benefits associated with planted forests.

Recommended action 57 Undertake studies on the short-, medium- and long-term trends in demand for all forest goods and services in the local, national and international markets, so that forest production can be more closely targeted to meet the diverse and changing market and community demands.

Recommended action 58 Intensify local, regional, national and international marketing efforts in order to realize the highest possible value returns from the forest products and promote improved utilization of the resources from sustainably managed forests.

5.2 FOREST MONITORING, GROWTH AND YIELD REGULATION

5.2.1 Integrated resource inventories

Principle 54 Integrated resource inventories are required to provide information on:
* the health of the trees, the forest ecosystem and the forest environment, any serious risk and damage factors;
* the state and potential development of biodiversity;
* the opportunities for wildlife conservation and management;
* the volume assortment of timber types and sizes, and the quality of the timber growing stock suitably qualified by species and management unit;
* the opportunities for outdoor recreation and the production potential of other non-wood forest products and values.

Principle 55 Forest inventories should obtain information on existing plans affecting land use, land and infrastructural development, land allocation and statutory and customary rights which may affect forest management and forest production. Wherever feasible, this information should be compiled in a Geographic Information System (GIS) to facilitate easy access, retrieval and evaluation of information – see Appendix 2. Such access is especially important at the regional land use level.

Principle 56 Easy access to comprehensive information is necessary for rational planning and forecasting, as well as for early adjustments of production forecasts, marketing strategies and management practices, so that management can remain realistic and relevant to community and market demands in a rapidly changing world.

Recommended action 59 Carry out continuous integrated forest inventories to determine where appropriate to local conditions:
* the overall health and compatibility of the social and political environment including the impact from encroachment, illicit felling or other kinds of product collection, deliberately lit fires, as well as national disasters;

- the progress in the development of the forest estate;
- the production status of both timber and non-timber products;
- the status of forest wildlife and game species;
- the trends in recreational and other forms of community use.

Recommended action 60 Establish and regularly monitor a comprehensive network of permanent inventory plots for volume assessment and collection of other necessary information and data.

Recommended action 61 Prepare and progressively update management information maps and resource inventory summaries to provide a sound basis for on-going community consultation and the regular adjustment of forest management plans.

Recommended action 62 Where feasible, take advantage of modern geographic information systems technologies to develop efficient and flexible data storage, retrieval, evaluation and forecasting systems.

5.2.2 Timber production

Principle 57 To achieve a sustained production of timber from each forest management unit, a reliable method for monitoring growing stock condition and increment is required. Similarly, a reliable system controlling timber allocations is also indispensable. Suitable systems of growth and yield prediction by means of simulation should be applied, and where needed be developed, to allow forest managers to respond to changing community and market demands in a manner consistent with the overall objective of sustainable production.

Recommended action 63 Regularly review Annual Allowable Cut (AAC) estimates in order to take account of deviations from predictions as a result of environmental changes, changes in crop or soil conditions or changes in demand for various forest products. Where appropriate, take advantage of recent developments in system modelling to simulate growth of forest stand and development of the forest enterprise. The objective is to ensure efficient, responsive, yet environmentally, economically and socially sound stand management, yield regulation and harvesting scheduling.

Recommended action 64 Use AAC estimations to set maximum production limits, but allow forest managers to adapt annual production according to current changes in market opportunities or community demands if necessary.

5.3 SILVICULTURAL OPERATIONS

5.3.1 Restoration and maintenance of soil fertility

Principle 58 The restoration and maintenance of soil fertility is just as important for the long-term management of planted forests as it was for their establishment. It is also fundamentally important for the sustainable management of future rotations established by replanting, coppicing or conversion to more complex types of mixed forests.

Recommended action 65 Review the Principles and Recommended Actions outlined in Section 4.1.2, and apply as appropriate during the post-establishment management phase, with emphasis on artificially or spontaneously developing mixtures of tree species and soil-cover crops.

Recommended action 66 Take particular care to manage the impact of harvesting activities by reviewing the procedures suggested for roads and site protection in Section 4.2.2.

5.3.2 Tending operations and weed control

Principle 59 While the competition effects of weeds and other non-crop forms of vegetation is generally less critical later in the forest rotation than in the establishment phase, tending and weed control may still be necessary to facilitate access for fire control, harvesting and other management activities. However, as discussed in Section 4.2.6, it is important to approach weed control in a considered manner to ensure that such actions are environmentally and ecologically sound and cost effective.

> *Recommended action* 67 Review Principles 30–32 and Recommended Actions·27–31 in Section 4.1.2 to develop tending practices appropriate for the circumstances of particular planted forests.

5.3.3 Thinning and pruning

Principle 60 Thinning and pruning planted forests can have a substantial influence on the end use and profitability of their products. The periodicity and intensity of thinning is usually influenced by growing stock conditions, increment responses to stand density, the availability of markets for smaller sized logs and the market incentives for larger sized end-of-rotation logs. Pruning is only usually justified where the accumulated sum of both direct costs and opportunity costs associated with temporary reductions in growth increment after pruning are more than offset by the ultimate increased value of knot free timber.

> *Recommended action* 68 Carefully evaluate the desirability of incorporating thinning and pruning schedules into the management of planted forests. In making provision for these activities, give particular consideration to the timing of such activities so as to reduce costs and produce maximum advantage. The spacing of trees and the regulation of mixing of tree species and the treatment schedules throughout the life of the forest stand should be designed by calculating backward from the desired features of the target mature crop, applying appropriate stand growth models.

5.3.4 Roading

Principle 61 Good access is essential for all management activities in planted forests. However, inadequately designed, inadequately constructed or inadequately maintained roads can cause problems of access in crucial periods of high demand, increased costs and substantial on-site and off-site environmental damage.

Principle 62 Forest road and track alignments should be carefully planned and preferably located on stable soils, have appropriately compacted surfaces, be well drained and have sufficient exposure to sunlight to ensure rapid drying after rain. Good drainage is crucial and roads should be located accordingly. Effective drainage must be assured by appropriate road construction and maintenance. Drainage specifications should be designed to ensure that structures are capable of handling peak discharges and that they minimize erosion and sedimentation in adjacent streams. Bridges and culverts should be of adequate capacity and be kept clear of obstructions.

> *Recommended action* 69 Develop an integrated plan to link roading, fire trail and harvesting access requirements in a manner consistent with the protection of the site and the wider environment.

> *Recommended action* 70 Provide close management supervision of all road construction activities to ensure conformity with approved planning specifications. Regularly monitor the condition of all roads and drainage structures and ensure that maintenance schedules are themselves maintained.

5.4 FOREST PROTECTION

5.4.1 Control of access

Principle 63 Planted forest areas must be protected from activities that are incompatible with environmental protection and sustainable timber production, such as encroachment by cultivators, illegal wood cutters, and illicit litter collectors. Local communities are often most effective in controlling access, provided that they view the planted forest as a benefit to them, and are given the authority and the means for effective access control.

Recommended action 71 Control public access to roads where these roads lead solely to forest working areas. Reduce pressures for forest encroachment by integrating forest management into wider rural development strategies. For example, consideration could be given to the possibility of managing special buffer zones within and beyond the border of the planted forests, to help provide for the basic needs of the local population living near the forest and prevent illicit uses and encroachments. In the buffer zone, multiple-use should have priority.

5.4.2 Protection from fire

Principle 64 As discussed previously in Section 4.2.2, fire can be a serious threat to the productivity and environmental quality of any planted forest. Fire risks must therefore be taken seriously and be addressed by active fire management programmes.

Recommended action 72 Review Principles 38 and 39 and Recommended Action 39 in Section 4.2.2, and adapt appropriately to the post-establishment management phase. Give particular attention to the problem of debris management following thinning or other harvesting operations.

5.4.3 Pest, disease and fire management

Principle 65 Pest and disease outbreaks and fire can occur at any stage during the rotation. Managers need to be prepared with well thought out control strategies.

Recommended action 73 Review the Principles and Recommended Actions outlined in Section 4.2.7 and apply as appropriate to the post-establishment management phase. Give particular attention to cultural practices that maintain vigorous growth and to the problem of debris and slash management following thinning or other forms of forest harvesting. Apply washdown and other simple hygiene practices to all machinery entering the forest from other areas, and prevent oil spills.

Recommended action 74 Make adequate provision in the annual budget for fire management and by means of a contingency fund for fighting fire outbreaks. A simple standard operating procedure and instruction manual should be prepared which can be easily understood and carried out by lower-level staff to be used in fire management and fire fighting.

5.5 HARVESTING AND PLANNING OF THE SUBSEQUENT ROTATION

Principle 66 Planted forests are highly artificial and in many cases narrowly focused on maximizing single-product functions. Natural and semi-natural forests are more widely focused on multiple use and perform more functions of production and protection.

Recommendation action 75 Before the final harvest felling is made, the decision should be taken on the design of the following forest generation. Wherever possible and feasible, this second generation should be planned to become more complex and diverse in order to promote ecological stability and diversity of production and of non-productive multiple forest functions.

Appendix 2
Technical Examples

Appendix 2.1:
Example of Management Plan Contents from Budango Forest Reserve, Uganda (Fourth Edition, 1 July 1997 to 30 June 2007)

Table of Contents

Appendix 2.2: Implementing HCVF in Small Forests

Contributed by Dr S Jennings, ProForest

1 Introduction

The concept of High Conservation Value Forests (HCVFs) forms part of the requirements of the FSC standard (Principle 9) and is also increasingly being used outside FSC certification. A complex and technical issue, there is little help for managers, particularly of small forests, on how to implement HCVF. This Appendix attempts to provide straightforward guidance for small-forest managers who wish to implement HCVF. It begins by introducing the HCVF concept and discussing in general terms what the six types of High Conservation Value (HCV) are. This description is relevant to managers of all scales of forests.

The Appendix continues by providing guidance on how small-forest managers can work out whether their forest management unit (FMU) contains any HCVF or not, and lays out some basic options for the appropriate management of them. Section 3 gives some examples of the type of monitoring that is likely to be appropriate for small-forest managers. The final section discusses some of the issues involved in consultation with stakeholders about HVCF.

1.1 A SAFETY NET FOR THE MOST IMPORTANT FORESTS

All responsible forest managers give consideration to the conservation of biodiversity and protection of watercourses or archaeological sites or other important values that their forest contains. This is reflected in the requirements of forest management standards. However, extra safeguards may be required when any such value is of such outstanding or critical importance that inappropriate management could have potentially disastrous consequences. This is the basis of the concept of HCVFs.

The HCVF concept was initially developed by the Forest Stewardship Council (FSC) and published as Principle 9 of their Principles and Criteria in 1999, and aims to provide just such a safety net for the world's most important forests. Forest managers are required to *identify* any HCVs that occur within their individual FMUs, to *manage* them in order to maintain or enhance the values identified, to *monitor* the success of this management and to *consult* with stakeholders at all stages of this process.

Specific requirements for forest managers are outlined in FSC-endorsed national standards. However, many countries do not yet have a national FSC standard, and those that do exist rarely provide guidance for small-forest managers. Many forest managers therefore find themselves in the situation that they are obliged to implement Principle 9 and yet have little guidance on what to do. While the 'HCVF Toolkit'[1] outlines in general terms what forest managers should do, there is no specific guidance within this document for small-forest managers.

1.2 HOW IS HCVF IMPLEMENTED IN SMALL FORESTS?

HCVF encompasses biological values, ecosystem services, social and cultural values. As outlined in Part Four, Section 15.6, there are six types of High Conservation Value. It is first important to gain a better understanding of what the six types of HCV mean in practice, and why they are considered so important. It should be noted that HCVs 1 and 4 can potentially cover a wide range of different things or processes and so it is convenient to divide these into separate elements, each of which may need to be considered separately.

---

1 The 'HCVF Toolkit Part 3' provides guidance for forest managers wishing to implement HCVF in the absence of guidance from a national standard. It is available from www.proforest.net

This section discusses each of the six HCVs in turn, describing the rationale behind them. Some of them are more likely to occur in small forests than others. Guidance on how to deal with them is given in Section 2.

1.2.1 HCV 1: Globally, regionally or nationally significant concentrations of biodiversity values

This value is intended to include areas with extraordinary concentrations of species, including threatened or endangered species, endemics, unusual assemblages of ecological or taxonomic groups and extraordinary seasonal concentrations of species.

Any forest that contains the species identified as HCVs, or which contains habitat critical to the continued survival of these species, will be an HCVF. This will include forests with many species that are threatened or endangered or many endemic species (eg 'biodiversity hotspots'). Exceptionally, it may even be that a single species is considered important enough to be an HCV on its own.

However, there will be many forests that contain rare or endemic species that are not HCVFs because there is not a globally, regionally or nationally significant concentration. These forests should still be managed appropriately, but they are not HCVFs.

Since there is a range of ways in which biodiversity values can be identified, this value has been sub-divided into four elements:

- **HCV 1.1 Protected areas:** Protected areas perform many functions, including conserving biodiversity. Protected area networks are a cornerstone of the biodiversity conservation policies of most governments and many NGOs and the importance of them is recognized in the Convention on Biological Diversity (CBD). Although the processes of selecting areas for protection have varied greatly in different countries and at different times, many are nonetheless vital for conserving regional and global biodiversity values.
- **HCV 1.2 Threatened and endangered species:** One of the most important aspects of biodiversity value is the presence of threatened or endangered species. Forests that contain populations of threatened or endangered species are clearly more important for maintaining biodiversity values than those than do not, simply because these species are more vulnerable to continued habitat loss, hunting, disease and so forth.
- **HCV 1.3 Endemic species:** Endemic species are ones that are confined to a particular geographic area. When this area is restricted, then a species has particular importance for conservation. This is because restricted range increases the vulnerability of species to further loss of habitat, and at the same time the presence of concentrations of endemic species is indicative of unique evolutionary processes.
- **HCV 1.4 Critical temporal use:** Many species use a variety of habitats at different times or at different stages in their life history. These may be geographically separate or may be different ecosystems or habitats within the same region. The use may be seasonal or the habitat may be used only in extreme years, when, nevertheless, it is critical to the survival of the population. This component includes critical breeding sites, migration sites, migration routes or corridors, or forests that contain globally important seasonal concentrations of species. In temporate and boreal regions, these critical concentrations will often occur seasonally (eg winter feeding grounds or summer breeding sites), whereas in the tropics, the time of greatest use may depend more on the particular ecology of the species concerned (eg riverine forests within tropical dry forests may be seasonally critical habitat for many vertebrate species). This element is included to ensure the maintenance of important concentrations of species that use the forest only occasionally.

1.2.2 HCV 2: Globally, regionally or nationally significant large landscape level forests

This part of the HCVF definition aims to identify those forests that contain viable populations of most, if not all, naturally occurring species. It often also includes forests that contain important sub-populations of very wide-ranging species (eg wolverine, tiger, elephant). It includes forests where ecological processes and ecosystem functioning (eg natural disturbance regimes, forest succession, species distributions and abundance) are wholly or relatively unaffected by recent human activities. Such forests are necessarily large. Where forest ecosystems naturally form a landscape level mosaic with other vegetation types and where many species use both forest and non-forest ecosystems, then it may be decided that this value relates to the mosaic of natural vegetation and not just the forest.

Large landscape level forests are increasingly rare and continue to be threatened throughout the world, through processes such as deforestation, forest fragmentation and degradation. Nevertheless, the occurrence of large, natural forests differs greatly from country to country. In countries where there has been extensive forest conversion, there may be no forests that would be included under this HCV. Alternatively, forests that are capable of maintaining most or all species may be so few that they are already well known. However, some countries retain a relatively large proportion of forest cover. Patterns of historical and current use, as well as current threats, may have reduced the ability of these forests to support the natural array of species. In such cases, the forest may need to be assessed to determine how significant it is at a global, regional or national level.

It is also worth emphasizing that the forest considered under HCV 2 is not necessarily confined to a particular administrative unit (eg FMU). This is because several contiguous FMUs may together form a significant large landscape level forest. An individual can be an HCVF under HCV 2 if it is part of a significant large, landscape level forest.

1.2.3 HCV 3: Rare, threatened or endangered ecosystems

Some ecosystems are naturally rare, where the climatic or geological conditions necessary for their development are limited in extent. Recent processes, such as land conversion, may have decreased their extent even further. Examples include montane forests in eastern Africa, cloud forests in Central America or riverine forests in semi-arid regions of Africa.

Other ecosystems have become rare through recent human activity, such as conversion of natural ecosystems into agricultural or other land use. It is often these ecosystems that are the most at risk in the future.

This value is designed to ensure that threatened or endangered forest ecosystems or types are maintained. It includes forest types that were previously widespread or typical of large regions. It also includes rare associations of species, even when the constituent species may be widespread and not threatened. These include:

- associations (intact or not) that have always been rare (eg beach forests along the Philippine coast);
- forest ecosystems, even if heavily disturbed or degraded, which are now rare or greatly reduced, and where intact examples are very rare (eg Atlantic forests (mata atlantica) of Brazil).

In these cases, the HCV is the rare ecosystem itself, which may be all or part of any particular forest. Native forest ecosystems or species assemblages that are characteristic of a region but are not rare or endangered should not be considered HCVFs under this part of the definition.

1.2.4 HCV 4: Forest areas providing basic services of nature in critical situations

All forests provide some services of nature, such as watershed protection, stream flow regulation or erosion control. These services should always be maintained under good management, a fact reflected in the requirements of most forest management standards. The value can be considered an HCV if the consequence of a breakdown in these services would have a serious catastrophic or cumulative impact. For example, a forest that forms a large proportion of the catchment area of a river that has a high risk of destructive flooding downstream may be critical in preventing flooding and would be considered an HCVF. It is this type of situation that HCV 4 attempts to identify.

Since there is a range of separate ecosystem services, this value has been sub-divided into three elements:

- **HCV 4.1 Forests critical to water catchments:** Forests play an important role in preventing flooding, controlling stream flow regulation and water quality. Where a forest area constitutes a large proportion of a catchment, it may play a critical role in maintaining these functions. The greater the risk of flooding or drought or the greater the importance of water usage, the more likely it is that the forest is critical to maintaining these services and more likely that the forest is an HCVF.
- **HCV 4.2 Forests critical to erosion control:** A second basic service of nature that forests provide is terrain stability, including control of erosion, landslides, avalanches and downstream sedimentation. All areas can potentially suffer some degree of erosion, but often the extent or risk of these is very low or the consequences minor. In some cases, however, forests protect against erosion, landslides and avalanches in areas where the consequences, in terms of loss of productive land, damage to ecosystems or property, or loss of human life, are severe. In these cases, the ecosystem service provided by the forest is critical, and it is these that should be designated HCVFs.
- **HCV 4.3 Forests providing barriers to destructive fire:** Fire is a part of the natural dynamics of many forest ecosystems, such as boreal forests in Canada or eucalypt forests in Australia. However, forest fires, whether started by natural causes or by humans, can sometimes develop into destructive, uncontrolled fire that can be a serious risk to human life and property, economic activity, or to threatened ecosystems or species. An HCV under this element includes forest that naturally acts as a barrier to fire in areas that are prone to fire, where the consequences are potentially severe.

1.2.5 HCV 5: Forest areas fundamental to meeting basic needs of local communities

The definition of HCVFs recognizes that some forests are essential to human wellbeing. This value is designed to protect the basic subsistence and security of local communities that are dependent on forests – not only for 'forest-dwelling' communities, but also for any communities that get substantial and irreplaceable amounts of income, food or other benefits from the forest.

Employment, income and products are values that should be conserved if possible, without prejudice to other values and benefits. However, management of HCVFs does not imply excessive and unsustainable extraction of resources, even when communities are economically dependent on the forest. Nor do they include the excessive application of traditional practices, when these are degrading or destroying the forests and the other values present in the forest.

A forest may have HCV status if local communities obtain essential fuel, food, fodder, medicines, or building materials from the forest, without readily available alternatives. For example, forests in the Brazilian Amazon that are used by extractivist communities (such as rubber tappers) as the sole or main source of economic activity would be considered HCVF under HCV 5. In such cases, the HCV is specifically identified as one or more of these basic needs.

HCV 5 applies only to basic needs. For example, for a community that derives a large part of its protein from hunting and fishing in forests where there is no alternative and acceptable source of meat or fish, the forests would constitute an HCVF. Another forest, where people hunted largely for recreational purposes (even if they did eat their catch) and where they were not dependent upon hunting, would not constitute an HCVF. The following would *not* be considered HCVFs:

- forests providing resources that are useful but not fundamental to local communities;
- forests that provide resources that could readily be obtained elsewhere or that could be replaced by substitutes.

Over time, an HCV may increase or decline in importance, with changing community needs and changes in land use. A forest, which was previously only one of many sources of supply, may become the only, or basic fundamental source of fuelwood or other needs. Conversely, needs may decline and disappear with time. For example, a forest that protected a stream that provided the only source of water for drinking and other daily needs to a community would cease to be an HCVF if a tube-well was constructed that provided water of sufficient quality and quantity for the community.

HCV 5 is determined by actual reliance on the forest by communities (even when this reliance is only occasional, as in the case of forests providing food in times of famine), rather than a future or potential situation. For example, the government of a particular country may have a scheme to generate employment and income for rural communities. If this is not implemented for all communities, or if some members of certain communities are unable or unwilling to take advantage of this and are consequently still dependent on forests for some of their basic needs, then a forest can still be an HCVF.

1.2.6 HCV 6: Forest areas critical to local communities' traditional cultural identity

As well as being essential for subsistence and survival, forests can be critical to societies and communities for their cultural identity. This value is designed to protect the traditional culture of local communities where the forest is critical to their identity, thereby helping to maintain the cultural integrity of the community.

A forest may be designated an HCVF if it contains or provides values without which a local community would suffer an unacceptable cultural change and for which the community has no alternative. Examples of HCVF under this part of the definition would include:

- sacred groves in India, Borneo and Ghana;
- forests used to procure feathers of the argus pheasant used by Dayak communities in Borneo in headdresses for important ceremonies.

This should include both people living inside forest areas and those living adjacent to it, as well as any group that regularly visits the forest. For example, the Maasai people of East Africa are mainly involved in herding cattle on the plains. However, they use forest as an integral part of their initiation rites and so should be considered in any discussion of forest use.

2 Identifying and managing HCVs in small forests

This section comprises a table providing suggestions on what owners and managers of small forests can do to identify HCVs within their forest areas. One column has been left blank so forest managers can use it as a template for recording whether each HCV (or HCV element) is found within their forest, including information on how that decision was reached.

HCV	Guidance for small forest managers	Notes
HCV 1 Globally, regionally or nationally significant concentrations of biodiversity values	**HCV 1 will not usually apply to small forests.** This is because although rare species, etc may well be present, small forests are unlikely to contain 'significant concentrations' of them. However, the forest manager should still check: *HCV 1.1 Protected Areas:* If the FMU is part of a protected area, then management should be consistent with the management aims of the protected area. If the FMU borders with a protected area, then the management should follow the guidance given for HCV 2, with additional discussions with the protected area management used to identify any specific measures necessary to conserve species' habitat or ecosystems. *HCV 1.2–4 Threatened and endangered species, endemic species and critical temporal use:* This will normally only apply if the forest contains one or more critically endangered species. The main challenge will be working out which rare species are present. It will always be worthwhile contacting local wildlife societies, conservation organizations and universities to find out what species are likely to occur within the area (see Chapter 15 for more ideas). Management measures should focus on maintaining important habitat features, such as: • Maintain buffer zones around the nests of rare birds or dens of animals. • Ensure that some areas of the habitat of a rare animal or plant are totally protected. This may require some improvements to habitat, for example leaving dead or snagged trees. • Stop operations close to a rare bird or animal during the breeding season. • Ensure that rare trees or shrubs are not disturbed during harvesting. • Educate staff and local communities not to hunt rare animals or collect rare plants.	
HCV 2 Globally, regionally or nationally significant large landscape level forests	**HCV 2 will not usually apply to small forests.** The only exception is if the FMU is part of a large, contiguous block of forest that covers tens of thousands of hectares. You will probably already be aware if this is the case, if not, it can be worked out from maps showing forest cover. In this case, as long as the forest manager is not severely disrupting the ecosystem or obstructing the movement of animals between the FMU and the rest of the larger forest area, then there should be little problem. The forest manager should therefore:	

HCV	Guidance for small forest managers	Notes
	• Minimize the impact of harvesting operations. • Avoid replacing natural forest with plantations of exotic species (which is prohibited by FSC-based standards anyway). • Avoid fencing. • Minimize the amount of permanent roads. • Talk to neighbouring forest owners to encourage them to do likewise.	
HCV 3 Forest areas that are in or contain rare, threatened or endangered ecosystems	It will always be worthwhile contacting local wildlife societies, conservation organizations and universities to find out whether your FMU contains a rare, threatened or endangered ecosystem. You will need to be able to tell them the location of the forest, what the forest looks like and what are the main tree species present. If the FMU does contain a rare, threatened or endangered ecosystem, then simple management measures include: • leaving some parts of the area as 'set aside', maintaining the original ecosystem untouched; • minimizing the impact of harvesting operations; • avoiding replacing natural forest with plantations of exotic species (which is prohibited by FSC-based standards anyway); • where important features of the rare ecosystem are known, making these the focus for habitat improvements (for example leaving dead or snagged trees as nesting or breeding sites).	
HCV 4 Forest areas that provide basic services of nature in critical situations	HCV 4 will not normally apply to small forests. Even though a small forest may provide a basic service of nature, such as watershed protection, the impact of a small forest on that is likely to be limited. In this case, compliance with best management practices (such as reducing the impact of harvesting operations and complying with regulations governing riparian protection zones and steep slopes) should be sufficient to ensure that the environmental services provided by the FMU are not degraded. The exceptions will be when the FMU is one of the few forests left in the immediate area. In this case, the forest manager should check: *HCV 4.1 & 4.2 Forests critical to water catchments and erosion control:* if all or part of the FMU is designated in the highest category of 'watershed protection forest' or 'erosion protection forest' (or equivalent) in national regulations, then	

HCV	Guidance for small forest managers	Notes
	it might be considered HCVF. If this is the case, the forest manager should, as a minimum, comply with the management requirements contained in the relevant regulations and should think of other management measures above and beyond these (eg increased width of riparian zones, minimizing the disturbance to the forest caused by harvesting). *HCV 4.3 Forests providing barriers to destructive fire*: small forests are normally critical barriers to fire only when they are adjacent to human settlements or protected areas or if they are adjacent to or contain archaeological or religious sites (such as a temple). If this is the case: • Seek advice from the local forest department on forest management techniques to reduce the risk of fires starting or spreading. • Discuss and plan fire protection measures with neighbours to ensure a co-ordinated response to fire.	
HCV 5 Forest areas fundamental to meeting basic needs of local communities	**HCV 5 will not normally apply to small forests.** Local communities often use forest resources to help them meet their basic needs (eg for collecting medicinal plants, construction material, firewood). However, a *small* forest is unlikely to be the only (or even the major) source of products *fundamental* to meeting basic needs of the community. Where the small forest is used by the local community, this should be taken into account in normal management, but would not necessarily be treated as an HCV. However, forest managers should check the extent to which local communities use resources from the forest. This should be done through consultation with local communities as part of on-going management (see Section 15.2), which should quickly reveal how important the FMU is to them.	
HCV 6 Forest areas critical to local communities' traditional cultural identity	Some small forests are of great cultural importance to local communities (eg when they are or contain a sacred grove, a temple, or a sacred burial ground). Consultation with local communities will be necessary (see Section 15.2 and Chapter 19) and should quickly reveal how important the FMU is to them.	

3 Monitoring HCVs in small forests

In principle, monitoring HCVFs is no different from monitoring any other aspect of forest management. The basics are given in Chapter 17 with additional pointers for biodiversity and social monitoring given in Section 15.6 and Chapter 21 respectively. However, in small forests, it is particularly worth remembering that monitoring programmes should be as cheap and simple as possible. Some examples of HCVs are given to illustrate the types of monitoring that are appropriate in small forests.

3.1 EXAMPLE 1: MONITORING RARE BIRD SPECIES

The small forest contains a large number of threatened or endangered bird species (HCV 1). Monitoring the populations of these would require a great deal of time as well as people skilled in bird identification. A much better option would be to work out the basic habitat requirements of the different species, probably with the help of a biologist from a local wildlife society, conservation organization or university department. Much of this information is likely to be known already and just needs to be put together. This would result in certain habitat features being identified as important to many or all of the bird species, such as:

- large hollow trees used for nesting cavities;
- patches of undisturbed forest interior habitat;
- fig trees;
- riverine forest;
- standing dead or snagged trees.

Hunting might also be identified as a threat to some of the bird species. Retention of the habitat features and control of hunting could be incorporated into management activities. Monitoring could focus on:

- operational supervision and monitoring to check that management practices are successful in retaining (or increasing) these habitat features;
- periodic counts of the number (or area) of these habitat features within the forest;
- recording casual observations of some of the more easily identified species (this could be done in collaboration with forest workers or local communities who use the forest);
- informal or periodic checks on hunting activities in the forest to ensure rare species are not being hunted.

3.2 EXAMPLE 2: MONITORING A SOURCE OF DRINKING WATER

The forest contains the source of a stream that is the only supply of water for drinking and other daily needs to a local village (HCV 5). The management option would be to create and maintain a riparian buffer zone. Within a small forest, monitoring water flow and quality (eg suspended solids load) would probably be far too expensive. More practical options for a small forest might include:

- operational supervision and monitoring to check that the riparian buffer zone is maintained during harvesting operations;
- consulting with the local health worker about trends in water-related diseases;
- consulting with the community themselves on their own observations on the quality of water.

3.3 EXAMPLE 3: MONITORING A SACRED BURIAL SITE

The forest contains a small, traditional burial site that is sacred to the local community (HCV 6). Consultation with the local community would be an essential part of any management of the site and the community should contribute to decisions about how the forest around it is managed. Such consultation might result, for example, in a plan for protecting the forest within 50 m of the burial site itself and the access path to it. Monitoring of this HCVF could therefore include:

- operational supervision and monitoring to check the protected areas of forest are maintained during harvesting operations;
- monitoring by the community and consultation to seek their views on the adequacy of management.

4 Consulting with stakeholders

The processes of identification, management and monitoring HCVs all require a degree of consultation. Consultation and participation in decision-making are considered in Section 15.2 and Chapter 19, where some guidance for small-forest managers is also given – consultation on HCV issues should follow this guidance.

When drawing up a list of potential stakeholders, it is worth remembering that:

- Conservation organizations, local representatives from government departments responsible for wildlife issues and local universities are important stakeholders for HCVs 1–3.
- Local representatives of government forestry or environmental protection departments are important stakeholders for HCV 4.
- For HCVs 5 and 6, consultation should always include the local communities in question.

Consultation is not a one-off event, and you will need to contact stakeholders more than once. The consultation will therefore vary over time. For example, you might contact a wildlife biologist at a local university when you are working out whether your forest is likely to contain a significant concentration of rare species (HCV 1). At this stage, your questions are likely to include: 'What threatened or endangered species occur within the region?', 'What forest types do these species occur in?' and 'What are the major habitat requirements of these species?'. To be able to answer these questions, the biologist will need to know some information about your FMU, such as where it is, what the main tree species are, something about the history of logging and hunting, whether the landscape around the FMU is largely forested or not, and what animals you have already seen in the forest.

At a later stage, when you are trying to work out appropriate management practices for the HCV, conversations with the same stakeholder are likely to focus on what habitat features are important to the species, how you can maintain (or increase) these habitat features, how many (or how much) you should have within the FMU, what the major threats on the species are, and so on.

Clearly, you will have to communicate in different ways with different stakeholders. Whereas a conservation NGO might be able to answer the question 'Is it important that I have species x, y and z in my forest?', a member of a local community would probably find the question strange. However, a conversation with the same community member about what animal species he used to see commonly when he was young but that are now unusual, about changes in forest cover in the region or about what animals he sees when he is hunting could yield information that is just as useful.

Finally, it is important that you keep some record of all consultation. This should include the date of the conversation, who you spoke with and a summary of what was said.

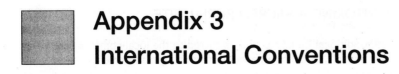

Appendix 3
International Conventions

Appendix 3.1: Chemical Pesticides Prohibited in FSC Certified Forests

From: Chemical pesticides in certified forests: interpretation of the FSC Principles and Criteria. FSC International Policy, FSC-IP-001, July 2002

Full policy and information on current derogations is available at www.fsc.org

Summary of policy

The main elements of the FSC policy on chemical pesticides are as follows:
Prohibited pesticides. The following categories of chemical pesticides are prohibited in forests covered by FSC-endorsed certificates.

- WHO Classes 1a and 1b. These are identified in Annex 1, and include aldicarb, parathion, oxydemeton-methyl, sodium cyanide and warfarin.
- Chlorinated hydrocarbons, including aldrin, DDT, dieldrin and lindane.
- Other persistent, toxic or accumulative pesticides, identified by their characteristics and defined thresholds.

Temporary exceptions apply, for cases identified in the policy and in national regulations and standards.
Other chemical pesticides may be used, at minimum levels and under the strict controls indicated by the P&C, regional standards and national regulations. Forest managers shall use pest control measures that have the least effect on the environment and people. They shall aim to control pests without using chemical pesticides, and demonstrate their efforts to achieve this aim. These aims should be clearly expressed in planning documents and operations. Phased reduction may be a good way towards achieving this goal. Regional standards will govern local interpretation in compliance with these guidelines, and may include case-by-case decisions on phasing out individual chemicals.

Chemical pesticides prohibited under the FSC rules of voluntary forest certification (listed in Annex 1 of the policy)

These chemicals are classed as 'prohibited' under Criterion 6.6 of the FSC Principles and Criteria for voluntary forest certification. Other chemicals are also prohibited if the active ingredients, or the formulations applied, comply with the definitions and characteristics shown in section 4 [of the full policy]. Temporary exceptions may be approved by the FSC board committee (Annex 4). Revised lists are published and circulated when necessary (see www.fsc.org).

Licensed chemical pesticides in formulations and mixtures, which do not contain these active ingredients, and do not exceed the thresholds in section 4 [of the full policy], are not immediately prohibited. They may be used, under careful controls, in accordance with the guidance included in the policy, and in accordance with FSC-endorsed Forest Stewardship Standards.

Name of chemical	Reason for prohibition	Temporary exceptions (updated April 2004)
aluminium phosphide	Toxicity similar to sodium cyanide. WHO Table 7.	
aldicarb	WHO Table 1, Class Ia.	
aldrin	CHC	
benomyl	Persistence: 6–12 months. Toxicity: LD50 100 mg/kg. LC50 60–140 mg/l. Mutagen	
brodifacoum	WHO Table 1, Class Ia.	Permitted for control of rodents in Chile, when they are vectors of Hanta virus transmission, in houses and camps. Permitted for control of Chacma baboons in Zimbabwe, subject to compliance with specified conditions. Temporary derogation (in Zimbabwe) granted until October 2004. See FSCPOL-30-601 Addendum 1 Brodifacoum derogation for details.
bromadialone	WHO Table 1, Class Ia.	Permitted for control of rodents in Chile, when they are vectors of Hanta virus transmission, in houses and camps.
carbaryl	Toxicity: LD50 of 100 mg/kg in mice.	
chlordane	Organochlorine. Persistence: half-life of 4 years. Toxicity: oral LD50 in rabbits approx 20–300 mg/kg.	
DDT	CHC	
diazinon	Toxicity: 0.0009 mg/kg/day. LD50 2.75–40.8 mg/kg.	
dicofol	Persistence: 60 days. Biomagnification: log Kow 4.28.	
dieldrin	CHC	
dienochlor	Organochlorine. Toxicity: LC50 of 50 mg/l in aquatic environments.	
difethialone	WHO Table 1, Class Ia.	Permitted for control of rodents in Chile, when they are vectors of Hanta virus transmission, in houses and camps.

Name of chemical	Reason for prohibition	Temporary exceptions (updated April 2004)
dimethoate	Toxicity: RfD 0.0002 mg/kg/day. LD50: 20 mg/kg in pheasants.	
endosulfan	Organochlorine. Toxicity: LD50 much less than 200 mg/kg in several mammals. RfD 0.00005 mg/kg/day.	
endrin	Organochlorine. Persistence: half-life >100 days. Toxicity: LD50 <200 mg/kg. Biomagnification high in fish.	
gamma-HCH, lindane	CHC	
heptachlor	Organochlorine. Persistence: half-life 250 days. Toxicity: LD50 100–220 mg/kg in rats, 30–68 mg/kg in mice. RfD 0.005 mg/kg/day. Biomagnification: log Kow 5.44.	
hexachlorobenzene	WHO Table 1, Class Ia.	
mancozeb	Toxicity: RfD 0.003 mg/kg/day.	
methoxychlor	Persistence: half-life 60 days. Toxicity: RfD 0.005 mg/kg/day. LC50 <0.020 mg/l for trout.	
metolachlor	Biomagnification: log Kow 3.45.	
mirex	Organochlorine. Persistence: half-life >100 days. Toxicity: LD50 50–5000 mg/kg. Carcinogen. Bioaccumulation high.	
oryzalin	Persistence: Half-life 20–128 days. Toxicity: LD50 100 mg/kg in birds.	
oxydemeton-methyl, metasystox	WHO Table 2, Class Ib.	
oxyfluorfen	Toxicity: RfD 0.003 mg/kg/day log Kow 4.47. (Goal, Koltar)	
paraquat	Persistence: > 1000 days. Toxicity: RfD 0.0045 mg/kg/day. Log Kow 4.47.	
parathion	WHO Table 1, Class Ia.	
pentachlorophenol	WHO Table 2, Class Ib.	
permethrin	Toxicity: log Kow 6.10. LC50 0.0125 mg/litre in rainbow trout.	Derogation to the end of 2003 for use with seedlings and young planted trees, when used with minimal impacts on insects and aquatic systems. (Permasect)
quintozene	Organochlorine. Persistence: 1–18 months. Toxicity: high. Biomagnification: log Kow 4.46.	

Name of chemical	Reason for prohibition	Temporary exceptions (updated April 2004)
simazine	Toxicity: RfD 0.005 mg/kg/day.	Permitted in state of Victoria, Australia, for residual pre-emergent control of grass and broadleaved weeds in eucalypt plantation establishment, subject to compliance with specific conditions. Temporary derogation granted until September 2006. See FSCPOL-30-601 Addendum 2 Simazine derogation for details.
sodium cyanide	WHO Table 2, Class Ib.	
sodium fluoroacetate, 1080	WHO Table 1, Class Ia.	Permitted for control of exotic mammals in Australia and New Zealand, where they cause damage to native plants or animals. Permitted in Western Australia, Victoria and South Australia for control of European Fox (as of April 2004; see FSCPOL-30-60. Addendum 3 Sodium fluoroacetate derogation for details).
2,4,5-T	Toxicity: medium to high in mammals. Often contaminated with dioxin.	
Organochlorine trifluralin	Toxicity: RfD 0.0075 mg/kg/day. Log Kow 5.07. LC50 0.02 mg/litre.	(under review, to be clarified)
toxaphene (camphechlor)	Organochlorine. Persistence: >100 days, high. Bioaccumulation high.	
warfarin	WHO Table 2, Class Ib.	Permitted for use against exotic mammal pests of native forests, including grey squirrels in UK, by approved operators with approved traps.
Pesticides containing lead (Pb), cadmium (Cd), arsenic (As), or mercury (Hg).		

Appendix 3.2: CITES – The Convention on International Trade in Endangered Species of Wild Fauna and Flora

CITES is an international convention designed to control the international trade in wild animal and plant species that are or might be threatened by trade. It first came into force in 1976; by 1997, 142 countries had signed the convention. Species are listed in three appendices to the convention:

- *Appendix* I lists endangered species which are largely banned from commercial international trade. There are special provisions allowing for trade in captive bred or artificially propagated animals or plants under certain circumstances.
- *Appendix* II lists species which may be traded internationally, but for which trade must be controlled and monitored through a system of permits. These are species which are not immediately endangered, but could become so by uncontrolled trade. It also includes 'look-alike' species whose trade needs to be controlled to assist with controlling similar Appendix I species.
- *Appendix* III is used by individual countries who seek assistance to control trade in native species which are not listed in the other two appendices.

Export permits are granted by the national CITES Management Authority, designated by the government of each signatory country; this is frequently the Ministry responsible for natural resources management.

There are approximately 298 species of plants included in Appendix I. Some of these are timber species, but most are species endangered by specimen collecting (sometimes in conjunction with other factors, such as habitat loss); for example orchids, cacti and cycads.

The tree species listed by CITES are (scientific and common names):

Appendix I

Abies guatemalensis	Guatemalan fir
Araucaria araucana	Monkey puzzle, Parana Pine
Balmea stormiae	Ayugue
Dalbergia nigra	Brazilian rosewood/Rio rosewood
Fitzroya cupressoides	Alerce/Chilean false larch
Pilgerodendron uviferum	
Podocarpus parlatorei	Parlatore's podocarp

Appendix II

Aquilaria malaccensis	Agarwood
Caryocar costaricense	
Guaiacum spp.	Lignum vitae/Tree of life
Oreomunnea pterocarpa	
Pericopsis elata	Afrormosia
Platymiscium pleiostachyum	Quira macawood
Podophyllum hexandrum	May-apple
Prunus africana	African cherry
Pterocarpus santilinus	Red sandalwood
Swietenia humilis	Central American/Mexican mahogany
Swietenia macrophylla	Big-leafed mahogany

| *Swietenia mahagoni* | Cuban mahogany/Caribbean mahogany |
| *Taxus wallichiana* | Himalayan yew |

Appendix III

Dipteryx panamensis	(Costa Rican populations)
Gonostylus spp.	Ramin
Podocarpus neriifolius	Yellow wood
Tetracentron sinense	

For further information on CITES contact: CITES Secretariat, 15 Chemin des Anemones, Case postale 456, CH-1219, Chatelaine, Geneva, Switzerland; or visit their website at: www.cites.org.

Additional information on trees of global conservation concern can be found at the UNEP-WCMC website: www.unep-wcmc.org

Appendix 3.3: Application of ILO Conventions to Forestry Operations

Summarized from the FSC policy paper 'FSC certification and the ILO conventions', May 2002. Available at www.fsc.org

The FSC policy of May 2002 is based on the premises that:

1 Forest managers are legally obliged to comply with all ILO Conventions which have been ratified in that country.
2 Forest managers are expected to comply with the eight ILO Core Conventions in all ILO member countries, by virtue of their membership, even if not all the conventions have been ratified.
3 FSC's policy for voluntary certification expects managers to comply with all conventions listed below, in all countries (including countries that are not ILO members, and that have not ratified the conventions).

The ILO Conventions with which compliance is required for FSC certification are:

29 Forced Labour Convention, 1930.
87 Freedom of Association and Protection of the Right to Organise Conventions, 1948.
97 Migration for Employment (Revised) Convention, 1949.
98 Right to Organise and Collective Bargaining Convention, 1949.
100 Equal Remuneration Convention, 1951.
105 Abolition of Forced Labour Convention, 1957.
111 Discrimination (Occupation and Employment) Convention, 1958.
131 Minimum Wage Fixing Convention, 1970.
138 Minimum Age Convention, 1973.
141 Rural Workers' Organizations Convention, 1975.
142 Human Resources Development Convention, 1975.
143 Migrant Workers (Supplementary Provisions) Convention, 1975.
155 Occupational Safety and Health Convention, 1981.
169 Indigenous and Tribal Peoples Convention, 1989.
182 Worst Forms of Child Labour Convention, 1999.
ILO Code of Practice on Safety and Health in Forestry Work (ILO 1998).
Recommendation 135 Minimum Wage Fixing Recommendation, 1970.
Conventions numbers 29, 87, 98, 100, 105, 111, 138 and 182 are Core ILO Standards covered by the 1998 ILO Declaration on Fundamental Principles and Rights at Work and its Follow-up.

ILO member states are expected to promote and realize these principles, even if they have not ratified the Conventions. The ILO Code of Practice is not a legal instrument, but it provides authoritative guidance on forest work.

Interpretation of the requirements of the ILO Conventions for Forest Managers

Poschen[1] provided an interpretation of the requirements of the ILO Conventions as they relate to FSC certification. These were incorporated into the FSC policy paper 'FSC certification and the ILO Conventions' and are summarized below.

CHILD LABOUR[2]

- Adherence to minimum age provisions of national labour laws and regulations, and of the international standards.
- No work under the age of 18 when it is likely to jeopardize health, safety or morals (unless there is special provision for safety, training or traditional community circumstances).

PROHIBITION OF FORCED LABOUR[3]

- No workers in debt bondage or other forms of forced labour engaged (including employees, self-employed or contractors).

INDIGENOUS AND TRIBAL PEOPLES[4]

- Communities have clear, credible and officially recognized evidence, endorsed by the communities themselves, of collective ownership and control of the lands they customarily own or otherwise occupy or use.
- Every reasonable effort is made to resolve conflicts through consultation aiming at achieving agreement or consent.
- The communities concerned have identified themselves as indigenous or tribal.
- The indigenous peoples are consulted through appropriate procedures and in particular through their representative institutions.
- Timely information about the proposed forest operations is provided giving details of expected impacts, benefit-sharing arrangements and decision-making procedures.
- Free and informed consent for forest management operations, if delegated by the indigenous peoples through their representative institutions, is freely expressed without coercion or duress.
- Cultural and traditional values are respected.
- Traditional access for subsistence uses and traditional activities is granted.
- Rights of local communities to natural resources pertaining to their land are respected and communities participate in the use, management and conservation of the resources. (Note: it is assumed that traditional uses are on a scale that does not threaten the integrity of the resources or the management objective.)
- Local and forest-dependent people have equal access to employment and training opportunities.
- All interested parties have access to relevant information [about management planning and operations].
- All interested parties have the opportunity to affect decision-making.
- Every reasonable effort is made to resolve conflicts through consultation aiming at achieving agreement or consent.

EQUALITY OF OPPORTUNITY AND TREATMENT[5]

- Employees are not discriminated in hiring, advancement, dismissal, remuneration and employment-related social security.

---

1 P Poschen, 'Social criteria and indicators for sustainable forest management. A guide to ILO texts'. Available at www.gtz.de/forest_certification under Materials and Working Papers, 2001
2 ILO Conventions 138 & 182, ILO Declaration 1998, or equivalent national legislation.
3 ILO Conventions 28 & 105, ILO Declaration 1998, or equivalent national legislation.
4 ILO Convention 169, or equivalent national legislation.
5 ILO Conventions 100 & 111, ILO Declaration 1998, or equivalent national legislation.

FAIR REMUNERATION[6]

- Wages or income of self-employed or contractors are at least as high as those in comparable occupations in the same region and in no case lower than the established minimum wage.

HEALTH AND SAFETY[7]

- A safety and health policy and a management system are in place which systematically identify hazards and preventative measures and ensure these are taken in the operations.
- All necessary equipment, tools, machines and substances are available at the worksite and in safe and serviceable condition.
- Safety and health requirements are taken into account in the planning, organization and supervision of operations.
- Where workers stay in camps, conditions for accommodation and nutrition comply at least with ILO Code of Practice on Safety and Health in Forestry.

RIGHT TO ORGANIZE AND BARGAIN COLLECTIVELY[8]

- All workers are able to form and join a trade union of their choice without fear of intimidation or reprisal.
- Collective bargaining with representative trade unions is carried out in good faith and with best efforts to come to an agreement.

RIGHT TO ORGANIZE AND DEFEND INTERESTS COLLECTIVELY[9]

- All interested individuals are able to form and join organizations of their choice without fear of intimidation or reprisal, and are well informed of their rights under these standards.
- Organizations of interested parties are accepted as participants in decision-making.

TRAINING[10]

- Managers and supervisors are in possession of an appropriate qualification, preferably one that is nationally recognized, ensuring that they are able to plan and organize forest operations and other elements of the management plan.
- All workers, as well as contractors and their workers and self-employed persons, are sufficiently educated and trained in the tasks they are assigned to and hold the relevant skill certificates.
- Policies and procedures make qualifications, skill and experience the basis for recruitment, placement, training and advancement of staff at all levels (without neglecting FSC Criterion 4.1).

---

6 ILO Convention 131, or equivalent national legislation.
7 ILO Convention 155, ILO Code of Practice on Safety (1988), or equivalent national legislation.
8 ILO Conventions 87 & 98, ILO Declaration 1998, or equivalent national legislation.
9 ILO Conventions 87, 98, 141 & 169, or equivalent national legislation or agreements.
10 ILO Convention 142, ILO Code of Practice on Safety (1988), or equivalent national legislation.

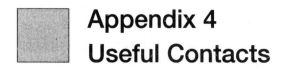

Appendix 4
Useful Contacts

Appendix 4.1 Internet Resources and Contacts

Centro Amzonico de Desarrollo Forestal (Cadefor)
Amazonic Center for Sustainable Forest Enterprise
Website: www.cadefor.org
E-mail: info@cadefor.org
Cadefor is a non-profit organization based in Santa Cruz, Bolivia. It aims to provide business management support, technical assistance and marketing communications support to the certified forest sector in Amazonian Bolivia. The Cadefor website (in Spanish and English) provides access to advice on a range of Amazonian wood species, including a database for species identification and wood properties and information about the Bolivian forestry sector.
Address: Calle 8 Este, No. 16, Equipetrol
      Santa Cruz de la Sierra, Bolivia
Tel: +591 3 34 24 353
Fax: +591 3 34 24 353

Center for International Forestry Research (CIFOR)
Website: www.cifor.cgiar.org
E-mail: cifor@cgiar.org
CIFOR is an international research institution committed to conserving forests and improving the livelihoods of people in the tropics.

   CIFOR produces a range of information and publications, many of which are available for free on its website. These include the CIFOR Criteria and Indicator Toolbox Series. CIFOR *News* is published in January, May and September every year. CIFOR News Online provides an advance view of the stories as they are completed.
Address: PO Box 6596, JKPWB
      Jakarta 10065
      Indonesia
Tel: +62 251 622 622
Fax: +62 251 622 100

Centro Agronómico Tropical de Investigación y Enseñanza (CATIE)
Website: www.catie.ac.cr
E-mail: comunicacion@catie.ac.cr or catie@catie.ac.cr
CATIE is a regional research centre and training centre for tropical agriculture and management of natural resources, dedicated to sustainable rural development and poverty reduction in tropical America. The CATIE website (in Spanish and English) offers a range of publications and forest management tools in Spanish and English.
Address: Apartado Postal 7170
      Turrialba
      Costa Rica
Tel: +506 556 6431
Fax: +506 556 1533

Centre for International Development and Training (CIDT)
Website: www.wlv.ac.uk/cidt
E-mail: cidt@wlv.ac.uk
CIDT is part of the University of Wolverhampton in the UK and provides training, consultancy and research in international development. CIDT offers a wide range of training courses in forest management, social forestry, biodiversity conservation and natural resource management (both in-country and in the UK).
Address: CIDT, University of Wolverhampton
                  Telford Campus
                  Telford TF2 9NT, UK
Tel: +44 (0)1902 32 32 19
Fax: +44 (0)1902 32 32 12

Collaborative Partnership on Forests (CPF)
Website: www.fao.org/forestry/site/2082/en
Email: unff@un.org
CPF is a partnership of 14 major forest-related international organizations, institutions and convention secretariats. It was established in April 2001, following the recommendation of the Economic and Social Council of the United Nations (ECOSOC). It aims to support the work of the UNFF and its member countries and to enhance cooperation and coordination on forest issues for the promotion of sustainable management of all types of forests. CPF produces a *Sourcebook on Funding for Sustainable Forest Management* www.fao.org/forestry/site/7148/en
Address: Department of Economic and Social Affairs
                  Secretariat of the United Nations Forum on Forests
                  DC2-2284, Two UN Plaza, New York, NY 10017, USA
Tel: +1 212 963 3160 / 3401
Fax: +1 917 367 3186

Commonwealth Forestry Association (CFA)
Website: www.cfa-international.org
E-mail: cfa@cfa-international.org
CFA produces the *International Forestry Review*, published four times a year and covering a range of aspects of forestry research. The *International Forestry Review* is available on-line: the full text of papers is available to CFA members and subscribers. Non-members can view abstracts free and the full text on a pay-per-view basis. The CFA website also provides links to a range of other forestry information sources.
Address: CFA, PO Box 142
                  Bicester OX26 6ZJ, UK
Tel: +44 (0)1865 82 09 35
Fax: +44 (0)1869 324 805

Conservation International (CI)
Website: www.conservation.org
Conservation International is a non-profit organization based in the USA, which aims to conserve natural heritage and biodiversity and to demonstrate that human societies can live harmoniously with nature. The website provides information on CI's programmes and conservation strategies. *Frontlines*, CI's quarterly newsletter is available free to download from the website, or can be ordered by e-mail (frontlines@conservation.org).
Address: 1919 M Street, NW Suite 600
                  Washington, DC 20036, USA
Tel: +1 202 912 1000

Department for International Development (DFID)

Website: www.dfid.gov.uk
E-mail: enquiry@dfid.gov.uk

DFID (previously ODA) is the UK Government's department whose aim is to reduce global poverty and promote sustainable development. DFID operates a number of funding schemes for individual scholarships and research projects. A large number of publications are available free of charge via the website, including many on environment, natural resources and land tenure.

Address: 1 Palace Street
London SW1E 5HE, UK
Tel (UK): 0845 300 4100
Tel
(Overseas): +44 (0)1355 84 31 32
Fax: +44 (0)1355 84 36 32

European Tropical Forest Research Network (ETFRN)

Website: www.etfrn.org
E-mail: etfrn@etfrn.org

ETFRN is a European research forum on tropical forestry. Although mainly aimed at researchers, ETFRN *News* covers current research issues related to tropical forestry. It can be downloaded free from the website.

Address: c/o Tropenbos International
PO Box 232, 6700 AE Wageningen
The Netherlands
Tel: +31 317 49 55 16
Fax +31 317 49 55 21

Food and Agriculture Organization of the United Nations (FAO)

Website: www.fao.org/forestry
E-mail: Addresses provided on website for individual departments and regional offices

The FAO website (in English, French and Spanish) offers a wide range of information on forest management, forest products and services, people and forests, environment and policy and institutions. Databases are available on-line including forestry contacts; short courses; education and research institutions; the FAO statistical database; forest valuation services; organizations working with NTFPs; forest genetic resources; and sources of funding to support SFM. A vast range of publications are available (some free, some for sale). *Unasylva*, FAO Forestry's international journal of forestry and forest industries, published quarterly in English, French and Spanish is available for free download on the website.

Address: Forestry Department, FAO
Viale delle Terme di Caracalla
00100 Rome, Italy

Forestales

Website: http://forestales.deamerica.net/

Forestales (Spanish) is an internet gateway to a large range of forestry-relevant websites for Spanish speakers.

Forest Certification Watch

Website: www.certificationwatch.org
E-mail: sfcw@sfcw.org

Forest Certification Watch provides up-to-date information about forest certification worldwide. FCW offers an e-mail newsletter, on-line news and conference information.

Address: PO Box 48122, Montreal
                QC, H2V 4S8 Canada
Tel: +1 514 273 5777
Fax: +1 514 277 4448

Forest Stewardship Council (FSC)

Website: www.fsc.org
E-mail: fsc@fsc.org

The FSC website (in English and Spanish) provides news and information about FSC certification. FSC policy and discussion documents are available to download for free. Up-to-date information is also available about accredited certification bodies, national initiatives and national/regional standards. Free subscription to FSC *News and Notes* is available via the website.

Address: FSC International
                Charles-de-Gaulle Str 5
                53113 Bonn, Germany
Tel: +49 228 367 66 0
Fax: +49 228 367 66 30

Forest Trends

Website: www.forest-trends.org
E-mail: info@forest-trends.org

Forest Trends is a non-profit organization which aims to promote incentives to diversify trade in the forest sector, moving beyond lumber and fibre to markets in a broader range of products and services. The website (in English, Spanish and partially in Chinese) offers Forest Trends' own *Technical Forestry Briefs* for forest policy-makers, as well as a comprehensive range of links to maps, databases and publications of other organizations.

Address: 1050 Potomac Street NW
                Washington, DC 20007, USA
Tel: +1 202 298 3000
Fax: +1 202 298 3014

Global Witness

Website: www.globalwitness.org/indexhome.html
Email: mail@globalwitness.org

Global Witness campaigns to highlight the link between the exploitation of natural resources, such as timber, and human rights abuses, particularly where the resources are used to fund and perpetuate conflict and corruption.

Address: PO Box 6042, London N19 5WP, UK
Tel: +44 (0)20 7272 6731
Fax: +44 (0)20 7272 9425

GTZ (Deutsche Gesellschaft für Technische Zusammenarbeit GmbH)

Website: www.gtz.de/english

GTZ carries out international development cooperation work in more than 130 countries, mainly on behalf of the German Federal Government. The GTZ website (in German and English) can be searched by sectoral theme, with pages on Forest Certification and Tropical Forests. Many publications are available to download free. The Forest Certification page contains a comprehensive range of Standards, Criteria and Indicators for SFM. Through the Forest Certification page there is a link to the Toolbox for Capacity Building in Forest Certification (also available at www.iac.wageningen-ur.nl/ForestCertification).

Address: Dag-Hammarskjöld-Weg 1-5
                 65760 Eschborn, Germany
Tel: +49 6196 79-0
Fax: +49 6196 79-1115

Instituto do Homem e Meio Ambiente da Amazônia (IMAZON)
(Amazon Institute of People and Environment)
Website: www.imazon.org.br
E-mail: imazon@imazon.org.br
IMAZON is a non-profit research institution promoting sustainable development in Amazonia, Brazil, through research, dissemination of information and capacity-building. The IMAZON website (in Portuguese and English) provides news, information about projects and some publications can be downloaded free.
Address: Caixa Postal 5101
                 Belem, PA, Brasil
                 CEP 666 13-397
Tel: +55 91 235 42 14
Fax: +55 91 235 01 22

International Association for Impact Assessment (IAIA)
Website: www.iaia.org
The IAIA website offer a number of Guidelines and Principles documents available to download free including 'Guidelines and Principles for Social Impact Assessment' and 'Principles of Environmental Assessment Best Practice'.

International Institute for Environment and Development (IIED)
Website: www.iied.org
E-mail: info@iied.org
IIED is an independent, non-profit organization doing research and influencing practice on a range of environment and development issues. Many publications are available free to download from the website in English, Spanish, Portuguese and French. IIED Forestry and Land Use publication series include: Small and Medium Forestry Enterprises; Instruments for Sustainable Private Sector Forestry; Policy that works for Forests and People; Forestry and Land Use; and Forest Participation Series. Other IIED publications are also available on a range of subjects including biodiversity and livelihoods, climate change and corporate responsibility.
Address: 3 Endsleigh Street
                 London WC1H 0DD, UK
Tel: +44 (0)207 388 2117
Fax: +44 (0)207 388 2826

International Labour Organization (ILO)
Website: www.ilo.org
E-mail: ilo@ilo.org
The ILO is an agency of the United Nations. It formulates international labour standards in the form of Conventions and Recommendations setting minimum standards of basic labour rights. The ILO website (in English, French and Spanish) has links with local sites in a number of other languages. Information is available on the ILO's conventions and their ratification. Publications are available to buy via the website; some (such as the Safety and Health in Forestry Work; an ILO Code of Practice) can be downloaded free.
Address: International Labour Office
                 4, route des Morillons

CH-1211 Geneva 22, Switzerland
Tel: +41 22 799 6111
Fax: +41 22 798 8685

International Tropical Timber Organization (ITTO)
Website: www.itto.or.jp
E-mail: itto@itto.or.jp
ITTO is an inter-governmental organization promoting the conservation and sustainable management, use and trade of tropical forest resources. The ITTO website (in English, French and Spanish) offers information about ITTO, and news about upcoming events worldwide. The ITTO newsletter, *Tropical Forest Update*, can be downloaded; free subscription to the print version is also available via the website. ITTO makes a range of its publications available for free downloading, including all ITTO Guidelines, Criteria and Indicators. The website provides links to a large number of inter-governmental, regional and civil society partner organizations.
Address: International Organizations Centre 5th Floor
Pacifico-Yokohama 220, 1-1-1 Minato-Mirai
Nishi-ku, Yokohama 220, Japan
Tel: +81 45 223 1110
Fax: +81 45 223 1111

IUCN (The World Conservation Union)
Website: www.iucn.org
E-mail: mail@iucn.org
IUCN is a membership organization made up of states, government agencies and NGOs. IUCN contributes policy advice on international conservation conventions, monitors the status of species on the IUCN Red List, and contributes to national biodiversity conservation plans. The IUCN Red List database can be searched on www.redlist.org by Red List category, country/region, habitat and major threat types. The IUCN Forest Conservation Programme page has a link to *Arborvitae*, the IUCN/WWF Forest Conservation Newsletter.
Address: Rue Mauverney 28
Gland, 1196, Switzerland
Tel: +41 22 999 0000
Fax: +41 22 999 0002

Metafore
Website: www.metafore.org
E-mail: info@metafore.org
Metafore is a non-profit organization that promotes business practices in the US that advance conservation, protection and restoration of forests worldwide. Metafore assists organizations seeking to improve the way they purchase and use forest products. The Resource Center provides tools for improved purchasing and trade of forest products.
Address: The Jean Vollum Natural Capital Center
721 NW Ninth Avenue Suite 300
Portland, Oregon 97209, USA.
Tel: +1 503 224 2205
Fax: +1 503 224 2216

MYRLIN (Methods of Yield Regulation with Limited Information)
Website: www.myrlin.org
Myrlin is a set of three simple software tools designed to assist in yield regulation for natural tropical forest. These include a stand table compilation module, a growth estimation tool and a harvesting

model. The tools are designed to work with limited knowledge of tree growth rates from locally applicable permanent sample plots. Each module tool comprises a Microsoft Excel workbook which can be downloaded free from the website.

Natural Resources Institute (NRI)
Website: www.nri.org
E-mail: nri@greenwich.ac.uk

NRI is an international, multi-disciplinary centre for research, consultancy and education for the management of natural and human resources. NRI publishes a range of books on sustainable development, which may be ordered from their website. Additionally, they also provide a number of on-line publications that may be downloaded for free. NRI offers undergraduate and postgraduate courses as well as research degrees. A variety of stand-alone short courses are also provided.

Address: University of Greenwich at Medway
                Central Avenue, Chatham Maritime
                Chatham, Kent ME4 4TB, UK
Tel: +44 (0)1634 880088
Fax: +44 (0)1634 880077

Overseas Development Institute (ODI) Forest Policy & Environment Group
Website: www.odifpeg.org.uk
E-mail: forestry@odi.org.uk

ODI is an independent British think-tank, which focuses on international development. The ODI Forest Policy & Environment Group (FPEG) conducts research on tropical forestry issues. The FPEG website (in English, French and Spanish) provides a broad range of forestry-related publications, including *Rural Development Forestry Network* (RDFN) papers, policy briefs and research papers, for free downloads.

Address: 111 Westminster Bridge Road
                London SE1 7JD, UK
Tel: +44 (0)20 7922 0300
Fax: +44 (0)20 7922 0399

ProForest
Website: www.proforest.net
E-mail: info@proforest.net

ProForest is an independent consultancy, which specializes in natural resource management and practical approaches to sustainability. The website (in English and French) provides details of the services ProForest offers as well as a list of ProForest publications, many of which are available for free.

ProForest offers an annual summer training programme located in Oxford, in addition to a variety of other training courses dealing with current issues in forest management, certification and sustainable natural resource management.

Address: 58 St Aldates
                Oxford OX1 1ST, UK
Tel: +44 (0)1865 243439
Fax: +44 (0)1865 790441

Programme for the Endorsement of Forest Certification Schemes (PEFC)

Website: www.perc.org
E-mail: pefc@pt.lu

PEFC is a non-profit NGO, founded in 1999, that provides a global umbrella organization for the mutual recognition of national forest certification schemes. The PEFC website provides access to the Pan-European Operating Level Guidelines (PEOLGs), PEFC reference and technical documents and links to national members' websites.

Address: PEFC Council ASBL
2éme Etage, 17 Rue des Girondins
Merl-Hollerich
L-1626 Luxembourg
Tel: +352 26 25 9059
Fax: +352 26 25 9258

Regional Community Forestry Training Center for Asia & the Pacific (RECOFTC)

Website: www.recoftc.org
E-mail: contact@recoftc.org

RECOFTC is an international organization that supports community forestry. RECOFTC designs and facilitates learning processes and systems that support the development of capacities of actors in community forestry. RECOFTC seeks to constructively promote dialogue between multi-stakeholders to ensure improved governance and equitable management of forest resources.

Address: PO Box 1111, Kasetsart University
Bangkok 10903, Thailand
Tel: +66-2 940-5700
Fax: +66-2 561-4880, 562-0960

Royal Institute for International Affairs (RIIA), Sustainable Development Programme

Website: www.illegal-logging.info

RIIA is an independent institute for analysis of international issues. The illegal logging website, funded by DFID and managed by RIIA's Sustainable Development Programme provides up-to-date information, briefings and documents on international initiatives to control illegal logging. An e-mail mailing list is also available.

Address: Chatham House
10 St James's Square
London SW1Y 4LE, UK
Tel: +44 (0)20 7957 5711
Fax: +44 (0)20 7957 5710

US Agency for International Development (USAID)

Website: www.usaid.gov

USAID is the US federal government agency working in development worldwide. The USAID website (in English and Spanish) provides access to the Sustainable Forest Products Global Alliance (also accessible at www.globalforestalliance.org), which aims to encourage responsible forestry practices and reduce illegal logging.

Address: Information Center, Ronald Reagan Building
Washington, DC 20523-1000, USA.
Tel: +1 202 712 4810
Fax: +1 202 216 35 24

Tropenbos International (TBI)
Website: www.tropenbos.nl
E-mail: tropenbos@tropenbos.org
Tropenbos International is an NGO that facilitates research and development programmes to meet the needs of forest policy-makers and forest users. TBI operates research sites in Cameroon, Colombia, Ghana, Indonesia and Vietnam. Cote d'Ivoire is a previous host to research sites. Website topics draw on practical research results and include reduced impact logging, SFM, NTFP Management and certification. The website (in English, Spanish and Dutch) offers the *Tropenbos Newsletter* and *Factsheets* free, on-line, as are numerous other Tropenbos documents. Other books and reports can be ordered via the website.
Address: PO Box 232, 6700 AE
Wageningen, Netherlands
Tel: +31 317 49 55 00
Fax: +31 317 49 55 20

World Bank
Website: www.worldbank.org
E-mail: eadvisor@worldbank.org
The World Bank's Revised Forest Strategy (2002) is outlined on the Forests and Forestry web pages and the full strategy can be downloaded. A great range of project documents and reports, publications and research working papers are available on the website.
Address: 1818 H Street, NW
Washington, DC 20433, USA
Tel: +1 202 473 1000
Fax: +1 202 477 6391

World Conservation Monitoring Centre (UNEP-WCMC)
Website: www.unep-wcmc.org
E-mail: info@unep-wcmc.org
The UNEP-WCMC provides information and databases related to conservation worldwide. The website provides access to the UNEP-WCMC species database for trees, which allows users to check the conservation status of tree species using their scientific names. There is also a link to the CITES listed species database (www.cites.org).

World Rainforest Movement
Website: www.wrm.org.uy
E-mail: wrm@wrm.org.uy
The World Rainforest Movement is an international network of citizens' groups of North and South involved in efforts to defend the world's rainforests. It works to secure the lands and livelihoods of forest peoples and supports their efforts to defend the forests from destructive developments. The website (in English and Spanish) provides news and campaign updates; monthly WRM electronic updates are available in English, Spanish, French and Portuguese.
Address: Maldonado 1858
Montevideo 11200, Uruguay
Tel: +598 2 413 2989
Fax: +598 2 418 0762

World Resources Institute (WRI)
Website: www.wri.org
E-mail: front@wri.org
WRI is an independent, non-profit, environmental research and policy organization. The WRI website offers access to a number of databases, downloadable data sets and maps, including information on forests, biodiversity and protected areas. A large number of publications are available to download free or hard copies can be bought via the website. The WRI website also links to their Global Forest Watch network (also available at www.globalforestwatch.org).
Address: 10 G Street NE (Suite 800)
                  Washington, DC 20002, USA
Tel: +1 202 729 7600
Fax: +1 202 729 7610

World Wide Fund for Nature (WWF)
Website: www.panda.org (with links to national organizations)
WWF is a global conservation NGO, working though advocacy and campaigning to highlight environmental issues and possible solutions. The Forests for Life campaign is one of WWF's six global issues. The WWF website offers a range of fact sheets, newsletters and publications on environmental and forestry issues. WWF also hosts the website of the **Global Forest and Trade Network** (GFTN) at www.panda.org/forestandtrade with information about forest management, certification and markets.
Address: Avenue du Mont Blanc, 27
                  CH-1196 Gland, Switzerland
Tel: +41 22 364 9111

Appendix 4.2: Further Reading

EIA AND ESIA

J Glasson, R Therivel, and A Chadwick, *Introduction to Environmental Impact Assessment: Principles and procedures, process, practice and prospects*. 2nd edition. UCL Press, London, 1999

P Morris, and R Therivel (eds), *Methods of Environmental Impact Assessment*. UCL Press, London, 2001

B Dalal-Clayton, and B Sadler, *The Status and Potential of Strategic Environmental Assessment*. IIED, London, 2004. Available at http://www.iied.org/spa/pubs.html#sea

FOREST OPERATIONS

G Applegate, F Putz and L Snook, *Who Pays for and Who Benefits from Improved Timber Harvesting Practices in the Tropics? Lessons learned and information gaps*. CIFOR, 2004. Available to download at www.cifor.cgiar.org

H C Dawkins and M S Philip, *Tropical Moist Forest Silviculture and Management: A history of success and failure*. CAB International, Wallingford, 359 pp,1997

D P Dykstra and R Heinrich, *FAO Model Code of Forest Harvesting Practice*. FAO, 1996. Available free at www.fao.org

T Enters, P B Durst, G B Applegate, P C S Kho and G Man, *Applying Reduced Impact Logging to Advance Sustainable Forest Management*. FAO Regional Office for Asia and the Pacific, Bangkok, Thailand, 2002. RAPP Publication 2002/14. Available free at www.fao.org Corporate Document Repository

FAO, *Code of Practice for Forest Harvesting in Asia-Pacific*. FAO regional office for Asia and the Pacific, Bangkok, Thailand, 1999. RAPP publication 1999/12. Available free at www.fao.org Corporate Document Repository

P van der Hout, *Reduced Impact Logging in the Tropical Rain Forest of Guyana: Ecological, economic and silvicultural consequences*. Tropenbos Guyana Series 6, 1999. Order from www.tropenbos.nl

ILO, *Safety and Health in Forestry Work: An ILO Code of Practice*. 1998. Available to buy, read on-line, or purchase in PDF format in English, French and Spanish from www.ilo.org

MONITORING AND FOREST RESOURCE ASSESSMENT

D Alder and T J Synnott, *Permanent Sample Plot Techniques for Mixed Tropical Forest*. Tropical Forestry Paper No. 25, Oxford Forestry Institute, 1992. Available to order from http://www.nhbs.com/services/oxforest.html

D Alder, *Growth Modelling for Mixed Tropical Forests*. Tropical Forestry Paper No. 30, Oxford Forestry Institute, 1995. Available to order from http://www.nhbs.com/services/oxforest.html

T B Boyle and B Boontawee (eds), *Measuring and Monitoring Biodiversity in Tropical and Temperate Forests: Proceedings of a IUFRO symposium held at Chiang Mai, Thailand August 27th–September 2nd 1994, Bogor, Indonesia*. CIFOR, 1995. Available to download from www.cifor.cgiar.org

J Carter (ed), *Recent Approaches to Participatory Forest Resource Assessment*. Rural Development Forestry Study Guide No. 2, Rural Development Forestry Network, ODI, 1996. Available to order from www.odi.org.uk/publications/order.html

M S Philip, *Measuring Trees and Forests*. 1994. Available to order from http://www.cabi.org/

J Wong, K Thornber and N Baker, *Resource Assessment of Non-wood Forest Products: Experience and biometric principles*. (English, French and Spanish). Non-Wood Forest Products 13, FAO, Rome, 128 pp, 2001. Order from www.fao.org Interactive Catalogue.

FOREST ECOLOGY, CONSERVATION AND HCVF

E F Bruenig, *Conservation and Management of Tropical Rainforests: An integrated approach to sustainability.* CAB International, Wallingford, 339 pp, 1996

R A Fimbel, A Grajal and J G Robinson (eds), *The Cutting Edge: Conserving wildlife in logged tropical forests.* Colombia University Press, New York, Chichester, West Sussex, 2001

S Jennings, R Nussbaum, N Judd and T Evans, *High Conservation Value Forests (HCVF) Toolkit; Parts 1-3.* ProForest, Oxford, 2004. Available at www.proforest.net

PLANTATIONS

M Garforth and J Mayers (eds), *Plantations, Privatization, Poverty and Power: Changing ownership and management of state forest plantations.* Earthscan, London, 2004. Order from www.earthscan.co.uk

C Cossalter and C Pye-Smith, *Fast-Wood Forestry – Myth and Realities.* CIFOR, Bogor, 2003. Available at www.cifor.cgiar.org

FORESTS, MARKETS AND LOCAL BENEFITS

IIED, *Instruments for Sustainable Private Sector Forestry.* A CD ROM containing six thematic reports, five country studies and an overview. IIED, London, 2004. Available also for download at http://www.iied.org/forestry/pubs/psf.html

N Landell-Mills and T I Porras, *Silver Bullet or Fools' Gold? A global review of markets for forest environmental services and their impact on the poor.* Instruments for Sustainable Private Sector Forestry series. IIED, London, 2002. Available at http://www.iied.org/forestry/pubs/psf.html

M Richards, J Davies and G Yaron, *Stakeholder Incentives in Participatory Forest Management: A manual for economic analysis.* ITDG Publishing, 256 pp, 2003. Order from www.itdgpublishing.org.uk

S Scherr, A White and D Kaimowitz, *A New Agenda for Forest Conservation and Poverty Reduction: Making markets work for low-income producers.* Forest Trends, Washington, DC, 2004. Available at www.forest-trends.org

S Shackleton, B Campbell, E Wollenberg and D Edmunds, *Devolution and Community-based Natural Resource Management: Creating space for local people to participate and benefit.* Natural Resource Perspective, ODI, No. 76, 2002. Available at www.odi.org.uk

FOREST POLICY AND GOVERNANCE

D Brown, K Schreckenberg, G Shepherd and A Wells, *Forestry as an Entry Point for Governance Reform.* ODI Forestry Briefing, No. 1, Overseas Development Institute, London, 2002. Available from www.odifpeg.org.uk

J Mayers and S Bass, *Policy that Works for Forests and People: Real prospects for governance and livelihoods.* Earthscan, London, 2004 (in press)

FOREST CERTIFICATION

S Bass, K Thornber, M Markopoulos, S Roberts and M Grieg-Gran, *Certification's Impacts on Forests, Stakeholders and Supply Chains.* IIED, London, 2001. Available at http://www.iied.org/forestry/pubs/psf.html

A Molnar, *Forest Certification and Communities: Looking forward to the next decade.* Forest Trends, Washington, DC, 2003. Available at www.forest-trends.org

R Nussbaum, *Group Certification for Forests: A practical guide.* ProForest, Oxford, 2002. Available at www.proforest.net

R Nussbaum, M Garforth, H Scrase and M Wenban-Smith, *An Analysis of Current FSC Accreditation, Certification and Standard-setting Procedures Identifying Elements which Cause Constraints for Small Forest Owners*. ProForest, Oxford, 2000. Available at www.proforest.net

R Nussbaum, S Jennings and M Garforth, *Assessing Forest Certification Schemes: A practical guide*. ProForest, Oxford, 2002. Available at www.proforest.net

R Nussbaum, and M Simula, *The Forest Certification Handbook*. 2nd Edition. Earthscan, London, 2004 (in press)

S Ozinga, *Footprints in the Forest: Current practice and future challenges in forest certification*. FERN, UK, 2004. Available at www.fern.org

M Richards, *Certification in Complex Socio-political Settings: Looking forward to the next decade*. Forest Trends, Washington, DC, 2004. Available at www.forest-trends.org

P Shanley, A Pierce, S Laird and A Guillén, *Tapping the Green Market: Certification and management of non-timber forest products*. Earthscan, London, 480 pp, 2002. Order from www.earthscan.co.uk

Glossary

active compartment	forest area (compartment) in which operations are on-going.
afforestation	new planting in areas not formerly covered in trees.
annual allowable cut (AAC)[1]	the volume of timber which may be harvested from a particular area of forest in any one year.
biodiversity[2]	(biological diversity) The variability among living organisms from all sources including terrestrial, marine and other aquatic ecosystems and the ecological complexes of which they are a part; this includes diversity within species, between species and of ecosystems.
biological control agent[2]	a living organism used to eliminate or regulate the population of other living organisms.
box cut	a road line cut or sunk into the ground, usually in order to reduce the gradient of a steep slope approaching a saddle or hill crest.
bucking[1]	the act or process of transversely cutting the stem or branches of a felled tree into logs. Also called cross-cutting.
buyers group	a group of suppliers and retailers whose members are committed to improving forest management around the world by purchasing their wood and wood products from well-managed forests. Part of the WWF Global Forest and Trade Network.
carbon offset	the result of any action specifically taken to remove from, and/or prevent the release of, carbon dioxide into the atmosphere in order to balance emissions taking place elsewhere.
certification body	an organization which conducts assessments of another organization's compliance with a set of accepted and agreed standards; certification bodies may be accredited by the standard-setting body, a separate accreditation authority.
chain of custody[2]	the channel through which products are distributed from their origin in the forest to their end use; usually used to imply a traceable route.
clear-felling[1]	a harvesting system in which all merchantable trees within a specified area of land are felled.
compartment	a sub-division of the forest management unit, usually scheduled to be harvested within a specified period of time.
protection area	an area permanently set aside from production in order to protect the ecosystem. Ideally protection areas should be linked by corridors of undisturbed forest to ensure they do not become isolated islands within a large area of harvested forest. For practical purposes, within an FMU, such protected areas can be defined as contiguous areas of more than 1 ha, but will often be a few hundred hectares in size.
contingency plan[3]	a plan designed to take account of a possible future event or circumstance (for example, action to take in the event of a chemical spill).
customary rights[2]	rights which result from a long series of habitual or customary actions, constantly repeated, which have, by such repetition and by uninterrupted acquiescence, acquired the force of a law within a geographical or sociological unit.

cutting licence	a legal permit allowing forest management or harvesting activities to occur on state land.
dbh[1]	diameter at breast height: the diameter of a tree measured at 1.3 m above ground level on the uphill side of the tree. For trees with large buttresses, a point above the main flare of the buttresses is often used for the measurement of diameter.
ecology[5]	the study of the inter-relationships between living organisms and their environment.
eco-label[3]	a label identifying manufactured products that satisfy certain environmental conditions.
ecological landscape planning	planning of operations to take account of and mimic the natural patterns of ecosystems affecting, for example, rotation length, size of clear-cut or species mix.
ecosystem[5]	a community of organisms and their physical environment interacting as an ecological unit.
endangered species[2]	any species which is in danger of extinction throughout all or a significant portion of its range.
endemic[5]	native to, and restricted to, a particular geographical region.
enrichment planting	the practice of planting trees within a natural forest to supplement natural regeneration.
enumeration	a pre-harvest inventory which records all commercial trees above a certain size (diameter).
Environmental Impact Assessment (EIA)	a process designed to identify actual and potential environmental effects of operations in order to plan how to minimize or avoid adverse effects and maximize positive effects.
Environmental Management System (EMS)[4]	the part of the overall management system that includes organizational structure, planning activities, responsibilities, practices, procedures, processes and resources for developing, implementing, achieving, reviewing and maintaining the environmental policy.
environmental performance bond	a deposit paid to an agency such as the forestry department, which is only returned if management operations are carried out satisfactorily.
environmentally sensitive area	an area of land which is particularly susceptible to damage by forestry operations and where operations are either prohibited or restricted: (for example, wetlands, watersheds, streamside buffer zones, conservation zones, recreation areas, areas near human settlements, sites of special ecological significance, habitats of rare or endangered species).
erosion[1]	the action of natural atmospheric conditions on any material exposed to them. Generally used to mean the wearing away of soil by the physical and chemical action of water.
exotic species[2]	an introduced species not native or endemic to the area in question.
extraction[1]	the process of transporting logs from felling site to landing: generally done by skidding or yarding.
fauna	animals, including mammals, birds, fish and insects.
flora	plants, including trees, shrubs, herbs and fungi.
forest inventory	a statistical assessment of the quantity and quality of a forest resource. Forest inventory for forest management planning includes: • a single inventory (static sample) to provide information on the current growing stock and rates of growth; • a recurrent inventory (dynamic inventory) to monitor growth rates and other changes in the forest.
forest management	an area of forest under a single or common system of forest management.

unit (FMU)

forest mensuration	techniques used for measuring trees and forests for forest inventory, the commonest parameters being dbh, cross-sectional area, length or height, form or shape, taper, age, volume.
forest organization	the company, community, association or other body with rights to manage the FMU on whose behalf forest management is carried out.
forest operations	activities in the forest, such as harvesting and roading, carried out as part of forest management.
genetically modified organism	biological organism whose characteristics have been deliberately modified by artificial manipulation of genetic material; does not include products of vegetative propagation or tree breeding.
global warming	predicted increase in temperature worldwide due to increasing levels of 'greenhouse gases' in the atmosphere, especially carbon dioxide.
habitat[5]	the locality, site and particular type of local environment occupied by an organism.
habitat trees	trees retained at harvest beyond the normal felling age specifically to provide living space for animals and other plants.
High Conservation Value Forests	Forests that possess one or more of the following attributes:

High Conservation Value Forests (continued):

(a) forest areas containing globally, regionally or nationally significant: concentrations of biodiversity values (eg endemism, endangered species, refugia); and/or large landscape level forests, contained within, or containing the management unit, where viable populations of most if not all naturally occurring species exist in natural patterns of distribution and abundance;

(b) forest areas that are in or contain rare, threatened or endangered ecosystems;

(c) forest areas that provide basic services of nature in critical situations (eg watershed protection, erosion control);

(d) forest areas fundamental to meeting the basic needs of local communities (eg subsistence, health) and/or critical to local communities' traditional cultural identity (areas of cultural, ecological, economic or religious significance identified in cooperation with such local communities).

high-loading	a cable extraction system where suspended cables are used to convey logs to the landing. One end of the log is attached to an overhead cable and lifted off the ground; the lower end of the log drags along the ground.
indigenous peoples[2]	'the existing descendants of the peoples who inhabited the present territory of a country wholly or partially at the time when persons of a different culture or ethnic origin arrived there from other parts of the world, overcame them and by conquest, settlement, or other means reduced them to a non-dominant or colonial situation; who today live more in conformity with their particular social, economic and cultural customs and traditions than with the institutions of the country of which they now form a part, under State structure which incorporates mainly the national, social and cultural characteristics of other segments of the population which are predominant' (working definition adopted by the UN Working Group on Indigenous Peoples).
integrated pest management (IPM)	a combination of preventative and curative procedures for controlling pests which aims to reduce reliance on chemical pesticides.
internal audit	an assessment carried out within the organization to ensure compliance

with its own procedures, standards and performance requirements.

keystone species
a species on which other species of animal or plant may depend for their survival. Keystone plant species are those which produce large fruit crops during periods of resource scarcity and are heavily relied upon by resident fruit and seed-eating animals.[6]

log landing[1]
a cleared area in which logs are collected during extraction in preparation for transport to the processing facility or other final destination. Also known as a log market or log yard.

monitoring
on-going assessment of performance and effects of management, including technical, environmental and social aspects.

natural forest[2]
forest areas where many of the principal characteristics and key elements of native ecosystems such as complexity, structure and diversity are present as defined by FSC-approved national and regional standards of forest management.

nutrient cycle
natural process where nutrients are taken up from the soil, used for growth in the plant, and are returned to the soil when the plant dies and decomposes.

native species[2]
a species that occurs naturally in the region; endemic to the area in question.

non-productive forest
areas on which forest management activities will not be carried out including conservation zones and commercially unproductive forest.

non-timber forest product (NTFP)[2]
all forest products except timber, including other materials obtained from trees such as resins and leaves, as well as any other plant and animal products.

nursery
an area designed for production of young plant stock.

peeler core
the central core of a log after it has been peeled for veneer or plywood, which becomes too small to peel further.

pioneer species
plants which are better adapted to growing in open situations and will rapidly colonize large gaps and disturbed forest.

plantation
an artificially planted forest area lacking most of the principal characteristics and key elements of a native ecosystem.

post-harvest assessment[1]
an evaluation undertaken to determine the degree to which a harvesting operation has met its stated objectives including compliance with environmental and social requirements.

producer group
A group made up primarily of forest owners and managers who are working towards or have already achieved forest certification, as well as processors and manufacturers that are working towards the exclusion of illegal timber from their supply chain and greater trade in certified forest products. Part of the WWF Global Forest and Trade Network (GFTN).

provenance
an identifiable region in the natural habitat of a species from where the seed of the trees (or their parents, etc) being used in a plantation originally came from.

reduced impact harvesting
harvesting procedures planned and carried out in a manner that minimizes environmental impacts.

regeneration
regrowth of the forest following harvesting, either through natural processes of seed dispersal and germination or artificially by planting.

residual stand
the forest which remains after harvesting and extraction.

secondary forest[2]
the ecosystems that regenerate from a substantial disturbance (flood, fire, land clearing or extensive and intensive logging) characterized by a scarcity of mature trees and an abundance of pioneer species and a dense understorey of saplings and herbaceous plants. Although secondary forests

frequently peak in terms of biomass accumulation well within one felling cycle, the transition to primary forests usually requires several rotation lengths, depending upon the severity of the original disturbance. Irreversible transformation of the underlying soil and nutrient cycle brought about by chronic or intense use may render it impossible for the original, primary forest type to return. (NB, new definition in development)

sediment[1]
the material products of erosion (soil, sand, clay, gravel and rocks) brought down watercourses and suspended in the water or deposited in outwash fans or on flood plains.

seed trees
mature trees of good form (both male and female) which are left to provide a seed crop for future harvests.

selection system
a silvicultural system in which some crop trees are retained during a particular felling entry; selection of those to be harvested and those to be retained may be based on dbh or other criteria.

side-casting
a technique used in road construction for building roads traversing steep slopes: material is cut from the upper side of the road and thrown down the slope to eventually provide a flat running surface.

silviculture[2]
the art of producing and tending a forest by manipulating its establishment, composition and growth to best fulfil the objectives of the owner. This may, or may not, include timber production.

skid trail[1]
the pathway over which logs are skidded in a ground-skidding extraction system. Also known as a snig track.

skidding[1]
terrain transport in which logs are dragged to the landing, rather than being suspended in the air or carried on a vehicle. Also known as ground skidding or snigging.

stakeholders
people who are interested in, or affected by, forest management and operations (for example, government agencies, local communities, employees, investors, environmental interest groups, customers and general public).

stock map
a large-scale map (usually 1:2000 to 1:5000) produced using information gathered during the pre-harvest enumeration and usually showing location of harvestable trees, planned extraction routes, topography, streamside buffer zones and other conservation zones.

streamside buffer zone
a strip of vegetation left undisturbed along the banks of a watercourse or lake, in which harvesting, extraction and (usually) establishment are prohibited, in order to protect the watercourse from sedimentation and damage. Also known as a riparian zone, river buffer zone, buffer strip.

succession[2]
progressive changes in species composition and forest community structure caused by natural processes over time.

sustained yield
production of forest products on a perpetual basis, ensuring that the rate of removal of forest products does not exceed the rate of replacement over the long term.

tenure[2]
socially defined agreements held by individuals or groups, recognized by legal statutes or customary practice, regarding the bundle of rights and duties of ownership, holding, access and/or usage of a particular land unit or the associated resources there within (such as individual trees, plant species, water, minerals).

threatened species[2]
any species which is likely to become endangered within the foreseeable future throughout all or a significant portion of its range.

topographic map
a map showing contour lines, which may also show watercourses, swampy

	areas, rock outcrops, gullies and other features.
tree-spotter	a forest technician with expert knowledge of tree species identification, often employed in inventory and forest survey work.
trucking time	the time it requires to transport logs from log landing to a delivery point, usually further processing or a shipping point.
use rights[2]	rights for the use of forest resources that can be defined by local custom, mutual agreements, or prescribed by other entities holding access rights. These rights may restrict the use of particular resources to specific levels of consumption or particular harvesting techniques.
yarding[1]	terrain transport in which logs are conveyed to the landing by cable or aerial systems that have the capability of fully or partially suspending the logs in the air during transit to the landing.
yield regulation	a technique for calculating and controlling the harvest level to ensure that sustained yield is implemented.

Notes

1 Definition from FAO Model Code of Forest Harvesting Practice
2 Definition from FSC Principles and Criteria glossary
3 Definition from Concise Oxford Dictionary 9th edn.
4 Definition from ISO definitions related to certification, labelling and EMSs
5 Definition from Dictionary of Ecology, Evolution and Systematics, 1982
6 ter Steege et al, *Ecology and Logging in a Tropical Rainforest in Guyana*, with Recommendations for Forest Management, Tropenbos Series 14, Tropenbos, Stichting, 1996

Index